Contents

(continued on next page)

building a self-sufficient life

Living a more independent lifestyle is a rewarding goal. By learning to do more things yourself, you gain control of the how and when and what of modern life.

The projects in this book are designed to help you reach the goal of self-sufficient living. The best part is that you don't need to sell your townhome and move to 80 acres in the mountains to make use of these projects. Pick off a few here and there, and you can participate in the self-sufficient way of being without making the sacrifices required when living off the grid.

Building a greenhouse is a great way for gardeners to jumpstart the growing season, or to introduce new, sensitive plants to your garden. See page 280.

The homestead may look very different today than in the pioneer days. But what we may lack in acreage, we can compensate for with creativity. Growing edibles in containers on a balcony is just one way to pursue your self-sufficient goals in an urban setting.

Home arts like weaving may seem obsolete, but participation in them is growing. If you really want to let your self-sufficiency flag fly, don't just learn to weave: build your own frame loom (see pages 236 to 241).

Hard or soft, apple or pear, homemade cider is pure goodness. But you just can't make it efficiently without a cider press. See pages 214 to 219 for an easy, DIY press you can make.

Why aspire to self-sufficiency? Because the pace of modern life can be overwhelming. All too often, the amazing technology that was supposed to free us and make life better and better instead becomes a drain on our money, time, and joy. As we've tried to improve our lives through progress and efficiency, we have ironically moved further and further from many of the things that make life healthy, rewarding, and fulfilling. It's just a matter of cause and effect. Medical advances that allow for gene therapy are used to create GMO foods of unproven safety. An ever-ready energy supply has ensured that our lives are more comfortable and convenient than ever before; but our environment pays for that convenience in the fallout from fracking, the pollution of coal-fired power plants, and the occasional devastating oil spill. All the while, modern life seems to demand that we move faster, do more, make more, and consume more. Along the way, we've become less able to do for ourselves, and more in the habit of buying everything we need regardless of the environmental and personal consequences.

And make no mistake, there are consequences. As we continue to fill landfills at an alarming rate, see our water tables and well water increasingly become contaminated, and learn more about how corporate farming is creating a less diverse food-supply chain, the global and environmental benefits of what's come to be known as "self-sufficient living" become more relevant and more apparent. By saving water, raising food that doesn't involve long-distance transport (and all the fuel and pollution that transportation entails), and finding new ways to create what we need using only the energy in our bodies, we're not only helping ourselves, we're helping the world.

Those upsides are attracting growing numbers of people to the self-sufficient lifestyle. The fact that you're reading this book means that, chances are, you're one of those people.

Slowly but surely, we're learning that quantity, speed, and convenience are not always equal to quality, wholesomeness, and satisfaction. People across the country are realizing that highly processed food, produced and packaged on an industrial level, is often not as truly good tasting or as good for you as simpler food grown in your own backyard. People are reconnecting with animal husbandry, from raising two or three chickens to a small herd of goats. They are finding the pleasure in working with living things that create resources, such as food and fertilizer, and rediscovering the simple joy of hand-woven textiles made on a loom they built themselves. They are realizing that there is a way to live that is gentler on the ecosystems of which we are so much a part. On a more pragmatic level, they are saving money with what they grow and produce and what they no longer need to buy.

Solar panels that collect and heat air to warm cold spaces in your home are mechanisms you can build yourself and duct in to your home's existing ductwork. Learn how on page 254.

But more than that, they are choosing to live better, more rewarding lives. They are rediscovering the simple pleasure of creating something of worth with their own two hands, of reconnecting with the garden and finding a sense of craftsmanship and purpose in everyday life. These people are creating new traditions for their families and getting reacquainted with timeworn, valuable experiences and skills. This is all part of choosing a richer life.

Thankfully, this movement is not an all-or-nothing proposition. Self-sufficiency doesn't necessarily mean that you need to move completely off the grid and live without modern conveniences or technology. People still need to go to work, attend family events, commute, and spend some down time doing nothing (or whatever they enjoy doing). Other people don't have the skills or inclination to build, much less use, something like a loom. All that's okay. Maybe you don't want to raise chickens, but building a cold

Raising city chickens is a booming business for today's urban homesteaders, as well as the hobby farmers working in a more rural setting. But keep in mind that for most chicken farmers, it's all about the eggs. We show you how to build your own coop (page 28) and brooder box (page 36).

Splitting firewood is great exercise and a good way to work off stress, but where will you store it? See pages 230 to 235 for a compact firewood shelter that's also attractive.

frame or adding a double-bin composter are more your speed. You can pick and choose; there's no evil in using the computer or watching a ball game on your big-screen TV. Self-sufficiency isn't about living in a cave. It's about taking steps to do for yourself, to move back toward simpler, more fulfilling practices in daily life, and to help the environment one step at a time. It's about living healthy and creating a healthier world. In short, it's about changing your life and the world around you in common-sense, achievable, and positive ways.

On that note, you should be realistic when you choose which projects in this book you'd like to tackle. If you're up for doing all of them, good for you. But it's just as good if you only feel comfortable tackling a few of them right now—or even just one. You can always do more later, and any step you take toward self-sufficiency is a step in the right direction. Even the most modest action is better than overreaching, getting burned out, and giving up on the notion of self-sufficiency altogether.

To give you as much choice as possible, we've included a wide range of projects that can help you lead a self-sufficient life. Learn about composting, planting, raising animals, preparing and drying food, constructing sheds and greenhouses, building fences—even assembling a wine cellar for your wines. The projects range from simple to involved, but none of them require anything more than moderate DIY and woodworking skills. We've tried to make the simple life as simple as possible to create.

No matter how much self-sufficiency you decide to implement in your own life, you'll be changing your life and the world. Do a little or a lot, and you can still feel satisfied that you're doing something. It may seem like a drop in the bucket, but that's exactly how buckets are filled.

Building your own chicken coop is a great project for DIYers of all skill levels (see page 28). Chickens provide nutritious eggs, natural garden pest-control, and organic fertilizer.

Even if you don't have a lot of space, you can use any sunny spot to grow produce in containers. A strawberry barrel like this can house up to 25 feet of strawberry plants and only takes up 2 square feet of space. To make your own barrel, see page 110.

Collected rainwater is perfect for watering the garden or the lawn. Rain is soft and free of most pollutants, so it is perfect for plant irrigation. Learn how to make your own rain barrels—flip to page 83.

Keeping bees is about more than fresh honey and beeswax. The threat of colony collapse is what is really driving the huge surge of interest in backyard apiaries. See pages 51 to 57 for information on building a Langstroth-style beehive for your home.

The Self-Sufficient Lifestyle

Self-sufficient living is a highly complementary practice—once you begin, you'll find that many parts of your home are connected, and that multiple systems of self-sufficiency contribute to one another, often corresponding with the natural cycles of the earth. Because of this interconnectedness, many of the projects in this book will naturally lead you to more and more projects that will help you maximize your self-sufficiency work.

For example, if you start a garden, the fruits and vegetables you grow will provide waste that will transform into the compost that will nurture next year's bounty. Setting up a rainwater collection system not only reduces your reliance on public utilities, but the fresh, soft water will also help your plants grow healthy. The hens you are raising for their eggs control garden pests and provide free fertilizer. At the end of the growing season, you'll likely be overwhelmed with vegetables and will need to find a means to store and preserve them—perhaps a basement root cellar could be a good option. And, by growing organic vegetables nurtured by compost and animal manure, you create a pesticide-free habitat for honeybees to prosper, while they, in turn, pollinate the plant life.

That said, you do not need to take on all the projects in this book at once. Start with the projects that naturally supplement the efforts your family is already making toward self-sufficiency. If you already recycle, a natural next step is to build compost bins and begin to make compost with food and paper waste as well. If you already maintain a beautiful flower garden, why not build a home for the honeybees that are already frequent visitors, allowing you to collect the honey they produce? If you already grow fruits and vegetables, why not build a solar fruit dryer or a drying rack, or, if you have apples, make a cider press? If you already garden, why not build a greenhouse?

For the newcomer, the projects on the following pages provide multiple opportunities to create a more self-reliant lifestyle. For the experienced self-sufficient homeowner, the step-by-step projects included here will provide you with the means to expand and streamline your efforts.

chickens & other creatures

A big part of self-sufficiency is producing your own food. But as delicious and nutritious as fruits and vegetables may be, protein is an important part of every diet. That's where food animals come in, especially chickens. Chickens provide an ongoing harvest of eggs and can supply valuable meat as well.

Beyond food, some living additions to your self-sufficient homestead provide support in different ways. Goats give you milk, which can be processed into cheese. Bees provide a valuable service in pollinating garden plants and supply delicious honey in the bargain— making them some of the most beneficial living things you can raise on your property. They also require very little care and upkeep, making them even more desirable for any homestead looking to become self-sufficient with a minimum of free time.

The projects in this section are all about taking the best care possible of the animals that call your home, well, home. These structures are specifically designed and suited to a given creature, but they have also been designed to look good. These are simple structures—you won't need homebuilding experience to put up the chicken coops we outline in the pages that follow, nor will you need to be a master woodworker to assemble your own honey-making beehive.

With a little bit of work and a modicum of tools and materials, you can make homes within your homestead for living things that will help you live more self-sufficiently.

chicken ark

01

Chickens adapt more easily to life on a small urban homestead than other farm animals. Consequently, raising urban chickens has grown in popularity in recent years. In fact, many municipalities have relaxed their restrictions and requirements when it comes to tending a small number of hens. It's worth noting that the same latitude has not been given to roosters—typically, you must file a document with the consent of all your neighbors before the city will let you keep a loud winged alarm clock in your yard.

Chickens can live comfortably in many different types of environments, from small urban backyards to roomy farms and ranches. They have a lot to offer the self-sufficient homeowner: high-quality meat, farm-fresh eggs, new chicks, and even nitrogen-rich garden fertilizer. Chickens do require a safe place to live, especially during the night. Usually, this takes the form of a chicken coop.

A chicken coop that can be easily moved around your yard or garden offers the added benefit of distributing natural fertilizer while the chickens feast on bugs and weeds in your yard. Some gardeners even design mobile coops to cap their raised beds so the birds can be moved around to fertilize the soil and prepare it for planting in the spring. The portable coop that is built on the following pages contains both a roost with laying nests and a protected scratching area. It is large enough to accommodate up to three full-size hens, is attractive enough to find a home in any urban, suburban, or estate yard, and is complete with all the essential components of a coop: space for roosting, nesting boxes for laying eggs, and easy access for the bird owner to refill food and water, clean the coop and replace bedding, and to collect eggs.

Build your chickens a safe and comfortable home that can easily be moved around your yard or garden. This type of portable coop, known as a chicken ark or a chicken tractor, should keep your birds safe from the elements and from predators.

Chickens are outdoor birds and prefer to roam within limits. They are natural pest-controllers and will stay healthiest if allowed to move about for a portion of every day.

Chickens spend most of their day foraging for bugs and tasty bits of greenery, then return to their coop every night, even if they've been wandering in the yard.

Approvals

The first step to starting your own chicken coop is to get permission from your local municipality. Many cities and towns allow homeowners to keep hens, but no roosters. Typically, there is also a limit on the number of hens you can keep, and the distance your coop must be located from your neighbors' windows. Check the regulations in your municipality as you develop your chicken ranching plan. It's also important to talk with your neighbors to seek out their consent, even if their written permission is not required.

Getting Started

If you plan to raise chickens for their eggs, you can buy them either as newborn chicks or as pullets (about four to six months old). Chickens usually start laying at about six months. However, the older the chickens are, the more they'll cost. Another point to keep in mind is that handling baby chicks when they're young makes it easier to handle them as adults—but also harder to butcher for meat. Do an online search for chicks for sale, or talk to local dealers for more information. Pullets do not require a special brood environment, as chicks do (see Brooder Box, page 36), but you should monitor their light exposure and heat when they're young. Keep your pullets in the coop for a week or so to help them get accustomed to their new home.

When they are old enough, allow your hens out of the coop during the day to peck and wander around a larger enclosed area, such as a small yard surrounded by a fence. Hens will not wander far. They love to dine on the bugs and weeds in your yard, and will produce a greater yield of healthier eggs if allowed to move around freely. At night, make sure all your hens are safely locked in to the coop to sleep.

Collecting Eggs

Collecting eggs from a brooding hen requires a careful hand and sound timing. Expect a good pecking if you reach into the nest while mother is awake. The best time to gently remove eggs from the nest is in the morning or during the night, when hens roost. This is also the best time to pick up a hen and move her, because she won't argue while she's sleeping. Eggs may be brown, white, or sometimes even light blue or speckled—depending on the breed of your chicken. No matter the appearance, what's inside will taste the same.

Gather eggs twice a day, and even more frequently during temperature extremes when eggs are vulnerable. The longer they sit in the nest, the more likely eggs are to suffer shell damage. After gathering, pat the eggs clean with a dry cloth. If they are noticeably dirty, wash them with warm water. Place clean, dry eggs in a carton and refrigerate.

Chicken Breeds

So, why do you want chickens?

Perhaps you dream of eggs with thick, sturdy sunshine yolks that are unbeatable for baking (and perhaps for selling at a farmers' market stand). Maybe you want to dress the dinner table with a fresh bird. Not sure? If you want both eggs and meat, you're safe with a dual-purpose breed such as the barred Plymouth Rock.

Next, consider the size flock you will need to fulfill your goals. This depends on land availability and how much produce you wish to gain. In other words, if volume of eggs or meat matters, then you increase your "production line." If your reason for raising chickens is to enjoy the company of a low-maintenance feathered pet—the meat and eggs are just a bonus—then a flock of three or four hens and possibly a rooster will get you started.

Layers

While all chickens produce eggs, laying breeds are more efficient at the job than other breeds; in short, layers lay more eggs. You can expect about 250 eggs per year or more if your layer is more ambitious than most. Laying hens tend to be high-strung, however, and while they lay many eggs, they show little interest in raising chicks. You may reconsider laying breeds if you want your hens to raise the next generation. Layers simply aren't interested—but they'll keep seconds coming to the breakfast table.

Meat Breeds

These chickens are classified based on size when butchered. Game hens weigh 1 to 3 pounds (0.5 to 1.4 kg), broilers (also called fryers) range from 4 to 5 pounds (1.8 to 2.3 kg), and roasters are usually 7 pounds (3.2 kg) or slightly more. You'll find cross-breeds ideal for the backyard, including broiler-roaster hybrids like the Cornish hen or the New Hampshire.

Dual-Purpose Breeds

Larger than layers but more productive (in the egg department) than meat breeds, dual-purpose breeds are the happy medium. Hens will sit on eggs until they hatch, so you can raise the next generation. There are many chickens that fall into this variety, and their temperaments vary. Many dual-purpose breeds are also heritage breeds, meaning they are no longer bred in mass for industry. They like to forage for worms and bugs, are known for disease resistance, and, essentially, are the endangered species of the chicken world.

Chickens raised for meat usually are purchased from a hatchery or a feed store when just a day or two old. Raising them to broiler weight (4 to 5 pounds) takes 6 to 8 weeks. During this time they will consume around 15 pounds of feed.

Ornamental chickens often make good pets. They enjoy human companionship. And they are a fun and visual addition to the yard!

Chickens do well in cold weather as long as they have a sheltered, insulated roosting area and their water supply is not allowed to freeze.

Building a Chicken Ark

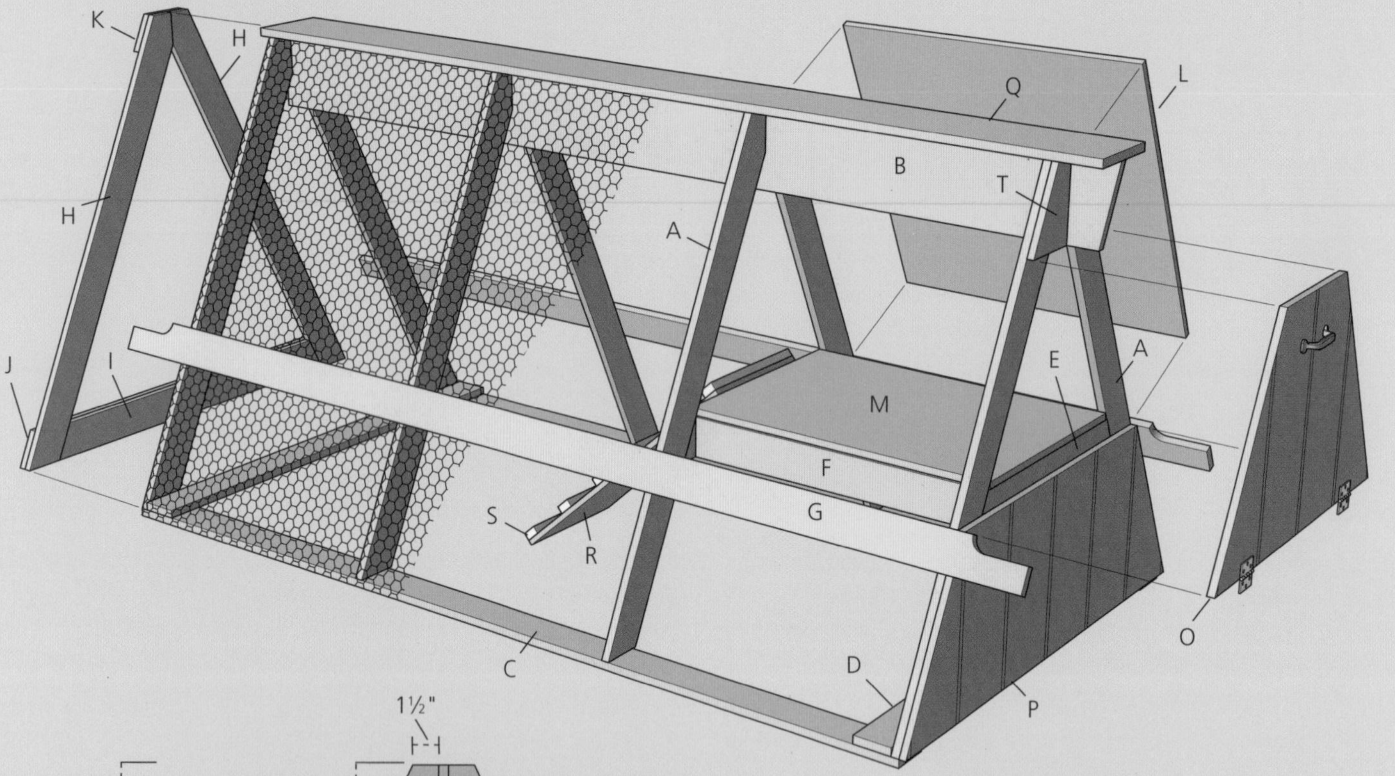
K
H
H
J
I
A
Q
L
B
T
E
A
M
F
G
S
R
O
C
D
P

Roost Wall Detail

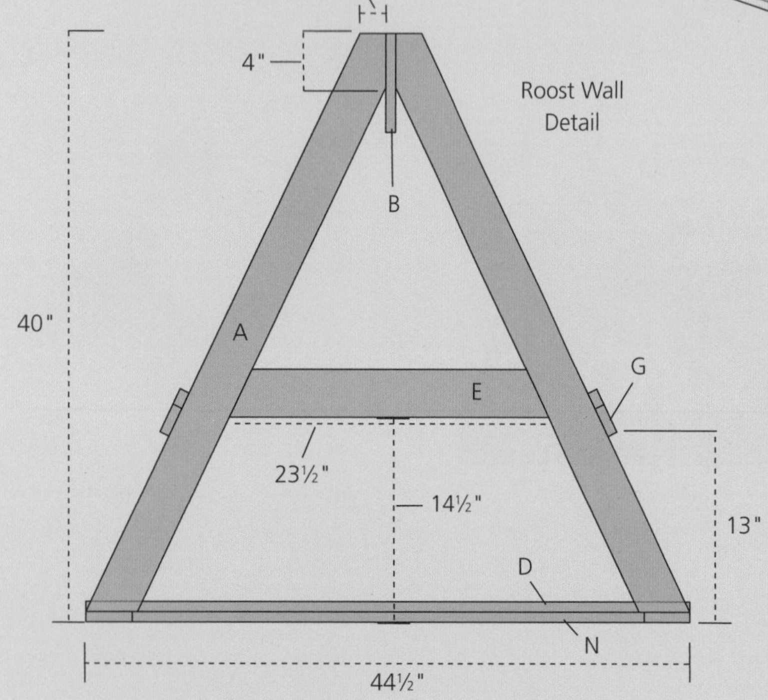
1½"
4"
B
40"
A
E
G
23½"
14½"
13"
D
N
44½"

Ramp Wall Detail

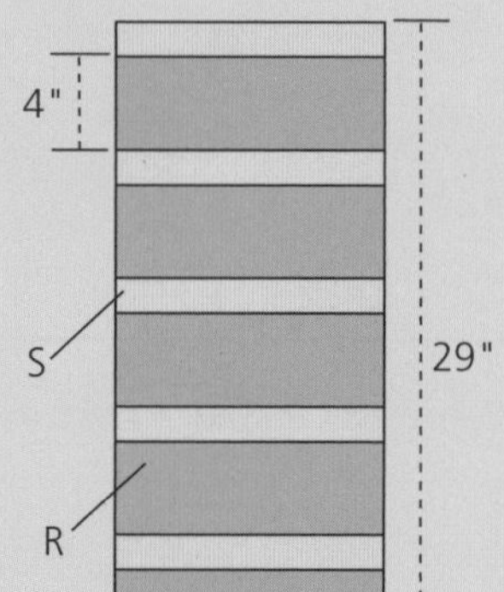
4"
S
R
29"

CUTTING LIST

Key	Part	Dimension	Pcs.	Material
A	Rafters	¾ × 3½ × 46"	8	1 × 4 pine
B	Ridge pole	¾ × 7¼ × 86"	1	1 × 8 pine
C	Base plate	¾ × 3½ × 81"	2	1 × 4 pine
D	Spreader	¾ × 3½ × 44½"	2	1 × 4 pine
E	Roost beam	¾ × 3½ × 23½"	2	1 × 4 pine
F	Roost joist	¾ × 3½ × 26"	2	1 × 4 pine
G	Handle	¾ × 3½ × 96"	2	1 × 4 pine
H	Door stile	¾ × 3½ × 47¼"	2	1 × 4 pine
I	Door rail	¾ × 3½ × 36¾"	1	1 × 4 pine
J	Door bottom brace	¾ × 3½ × 44½"	1	1 × 4 pine
K	Door gusset	¾ × 7¼ × 8⅛"	1	1 × 8 pine
L	Roost side	½ × 27¾ × 29"	2	Siding panel
M	Roost floor	½ × 20 × 27½"	1	Siding panel
N	Base filler	¾ × 3½ × 36¾"	1	1 × 4 pine
O	Roost door	½ × 19½ × 29½"	1	Siding panel
P	End wall	½ × 44⅝ × 16⅛"	1	Siding panel
Q	Ridge board	¾ × 5½ × 88"	1	1 × 6 pine
R	Ramp	½ × 12 × 29"	1	Siding panel
S	Ramp battens	¾ × 1½ × 12"	6	1 × 2 pine
T	Roost Door Filler	½ × 1½ × 4"	2	Sliding

TOOLS & MATERIALS

Circular saw

Jigsaw

Speed square

Tape measure

Power miter saw

Drill

Galvanized wood screws (1¼", 2", 3½")

Galvanized common nails

Hammer

Eye protection

Sander

Spacers

Galvanized finish nails

Pneumatic narrow crown stapler

Poultry netting (chicken wire)

Galvanized U-nails

Aviation snips

Pliers

Galvanized butt hinges (6)

Galvanized T-hinges (2)

Galvanized latches (4)

Door handle

Eye and ear protection

Work golves

How to Build a Chicken Ark

Make the eight rafters by laying out one rafter according to the diagram on page 22. Use this rafter as a template for marking and cutting the remaining rafters from 1 × 4 pine.

Cut an 8' 1 × 8 to 86" to make the ridge pole. Cut at a bevel of 10 to 15° for a decorative tail cut at the roost end. Then, attach the rafters on one side of the ridge pole with 2" deck screws driven through the ridge pole and into the rafters. The rafters should be spaced 26" apart in the field area.

Attach the rafters on the opposite side of the ridge pole by driving 3½" deck screws through counterbored pilot holes in the rafter tops and into the ridge pole.

Attach the base plates to the bottom ends of the rafters with deck screws. The outside edges of the base plates should be flush with the outside edges of the rafters. Then, attach the spreader at each end of the framework. Make sure the rafter legs are spaced consistently.

Attach the roost beams between the legs of the outside rafters. The bottoms of the beams should be 14½ " up from the bottom of the ark. **TIP:** Before cutting the beams to size, hold the workpiece against the rafters to make sure it will fit. Attach the beams with 3½" deck screws predrilled and driven up toenail style through the bottom edges of the beams and into the rafters.

Attach the roost joist boards between the beam boards, flush against the inside edges of the rafters. Drive three 2" deck screws per joint.

Make the floor board for the roost. Cut the panel to size and then test the fit. Install the floor with pneumatic staples or screws driven into the beams and joists.

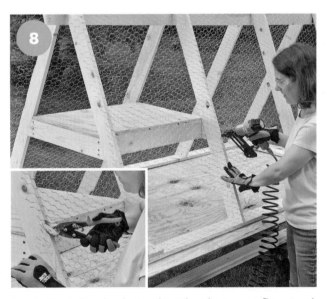

Cover the ark with galvanized metal poultry netting. Cut strips of netting to fit each side and then staple the netting with a pneumatic stapler and ⅞" narrow crown staples (otherwise, hand nail it with U-nails: A staple gun is inadequate for this job). Trim the ends of the poultry netting (inset).

(continued)

How to Build a Chicken Ark (continued)

Cut the profiles onto the ends of the handles. Make a cardboard template of the profile and then trace the profile onto each end. Cut with a jigsaw and then sand the cuts and edges smooth.

Attach the carrying handles for the ark with 2" deck screws driven into pilot holes. The bottoms of the handles should be about 13" up from the bottom of the ark—make sure they are parallel to the base. Take care when drilling the pilot holes to keep them centered in the edges of the rafters. Make sure the handle end overhangs are equal and that the profiles are pointing upward.

Attach the roost side panels to the ark frame with pneumatic nails or staples or with 1¼" deck screws.

Install the end wall and the hinged roost door. The wall is affixed permanently with screws or staples. The door should be hinged from below. Install snap latches to the roost door fillers to secure the door in place when it is raised.

Make the frame for the ark door that fits against the open end of the ark. Rather than hinges, use latch hardware to hold the door in place. Begin by making the triangular door frame. Then, attach poultry netting to the exterior face of the door.

Complete the ark door by cutting and attaching the door brace and the door gusset and then securing the door against the end of the ark frame with latch hardware. Consider adding a utility handle to the gusset for ease of door handling.

Cut a 1 × 6 ridge board to span from the top of the ark door gusset to the beveled end of the ridge pole at the opposite end of the ark. Center the ridge board side to side and nail it to the ridge pole to cover the gaps.

Make the ramp and attach 1 × 2 battens to create purchase for the birds. Tip the coop up on end, then attach the ramp to one side of the roost floor with butt hinges. **TIP:** Drill a hole in the end of the ramp, tie a rope to it, and thread the rope out through the top of the ark so you can use it to raise and lower the ramp as needed.

chicken coop

This chicken coop is just over 4 feet by 8 feet, with a 6-foot-wide roof, so it won't take up much space in your yard. It's big enough to comfortably house up to six chickens, and they'll stay dry and safe from predators under the large roof. Two people (or one strong person with a dolly) can easily roll the coop around the yard, and in the winter the plywood enclosure is tight enough to keep the chickens comfortable.

The framing for the coop has been keep to a minimum so that it's easier to move and quicker to build. We've also used 2 × 2s instead of 2 × 4s as much as possible to lighten the weight. Once you have all the materials on hand, you can build the coop in a day. If you have more chickens to house, the basic plan can easily be expanded—just keep the width and height the same and extend the length so that you can make economical use of 4 × 8 sheets of plywood.

The coop is large enough for two or three nesting boxes, which you access by opening the large door in front and leaning in. These can be as simple as plastic crates. The chickens will also appreciate a roost inside the coop. Make this from a closet rod pole or a length of 2 × 2, placing it about 18 inches up and running it from wall to wall in the middle of the coop. A window opening under the eave in the back provides ventilation, even in heavy rain. In the winter, simply cover it with plexiglas or plastic. You can also look into simply installing a small, inexpensive, utility-grade window in one of the sides—a bit luxurious for chickens, but it will give you an easy way to regulate the temperature and airflow inside the coop.

House your chickens in style in this large, movable coop. The chickens will appreciate the extra space, and the wide roof helps keep the ground from getting muddy.

Building a Chicken Coop

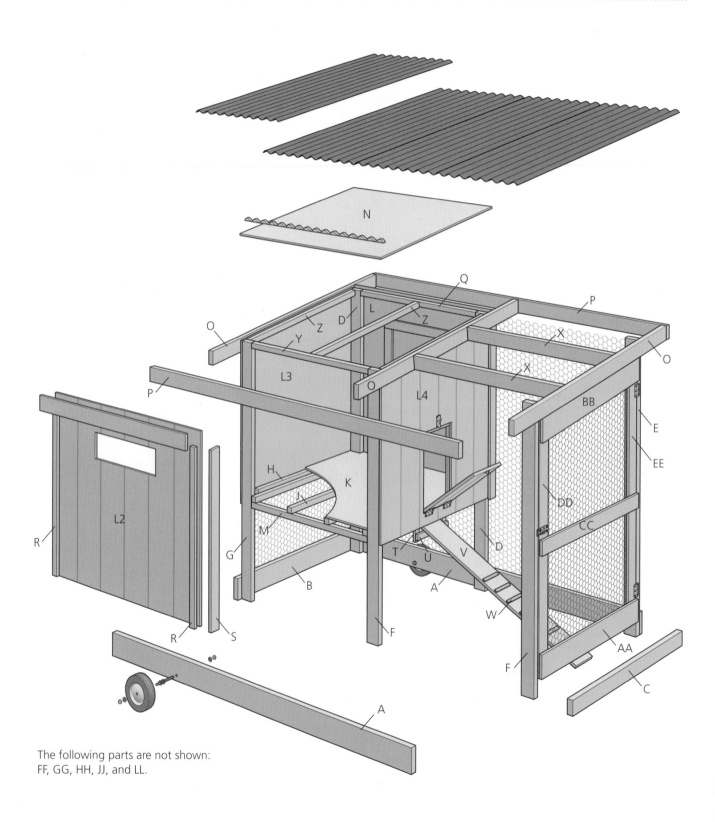

The following parts are not shown:
FF, GG, HH, JJ, and LL.

CUTTING LIST

Key	No.	Part	Dimension	Material
A	2	Base	1½ × 5½ × 93"	2 × 6 P.T.
B	1	Base end	1½ × 5½ × 50"	2 × 6 P.T.
C	1	Base door end	1½ × 3½ × 50"	2 × 4 P.T.
D	2	Front posts	1½ × 3½ × 72"	2 × 4 P.T.
E	1	Front door side post	1½ × 3½ × 72½"	2 × 4 P.T.
F	2	Rear posts	1½ × 3½ × 66"	2 × 4 P.T.
G	1	Rear door side post	1½ × 3½ × 66½"	2 × 4 P.T.
H	2	Floor joist support	1½ × 1½ × 39½"	2 × 2
J	4	Floor joist	1½ × 1½ × 46¾"	2 × 2
K	1	Floor	46¾ × 46¾"	½" plywood
L	1	Front wall	48 × 48 × ⅝"	T1-11 siding
L2	1	Rear wall	48 × 41¾ × ⅝"	T1-11 siding
L3	1	End wall	46¾ × 48 to 41¾ × ⅝"	T1-11 siding
L4	1	Door side end wall	46¾ × 48 to 41¾ × ⅝"	T1-11 siding
M	2	Floor rim joist	1½ × 1½ × 36 ½"	2 × 2
N	1	Roof	48 × 48	½" plywood
O	3	Roof beams	1½ × 3½ × 69"	2 × 4 P.T.
P	2	Fascia	1½ × 3½ × 96"	2 × 4 P.T.
Q	2	Top trim	¾ × 3½ × 48"	1 × 4
R	4	Corner trim	¾ × 1½ × 45" (varies)	1 × 2
S	4	Corner trim	¾ × 2½ × 45" (varies)	1 × 3
T	2	Ramp hanger	¾ × 1½ × 6"	1 × 2
U	1	Ramp support	¾ × 1½ × 13"	1 × 2
V	1	Ramp	¾ × 7¼ × 60"	1 × 8
W	14	Ramp battens	¼ × ¾ × 6"	Screen mold
X	2	Roof rafters	1½ × 3½ × 44"	2 × 4 P.T.
Y	2	Roof support	1½ × 1½ × 39½"	2 × 2
Z	3	Roof supports	1½ × 1½ × 44"	2 × 2
AA	1	Bottom rail	¾ × 5½ × 43¼"	1 × 6 cedar
BB	1	Top rail	¾ × 5½ × 43⅝" (angle to angle)	1 × 6 cedar
CC	1	Center rail	¾ × 3½ × 43¼"	1 × 4 cedar
DD	1	Handle stile	¾ × 3½ × 58¹¹⁄₁₆" (high edge)	1 × 4 cedar
EE	1	Hinge stile	¾ × 3½ × 64¼" (high edge)	1 × 4 cedar
FF	1	Door stop	¾ × 1½ × 42"	1 × 2
GG	2	Entry door stop	¾ × 2½ × 35½"	1 × 3
HH	1	Entry door stop header	¾ × 2½ × 24"	1 × 3
JJ	2	Ramp door stop	¾ × 2½ × 19½"	1 × 3
LL	1	Ramp door stop header	¾ × 2½ × 16¼"	1 × 2

TOOLS & MATERIALS

Miter saw
Circular saw
Jigsaw
Drill
Hammer
Stapler
Wrench
Framing square
Countersink bit
½" wood bit
Deck screws—1¼", 1⅝", 3"
2½" self-tapping screws or deck screws
(12) ⁵⁄₁₆ × 3½" galvanized carriage bolts
(2) 5½" × ½" galvanized hex bolts with 4 nuts and 8 washers (for tires)
(4) 1½" galvanized metal angles
9 × 1" roofing screws

(8) ⅜" × 4" exterior, self-tapping lag screws
4d galvanized casing nails
(3 pairs) 3" × 3" exterior grade hinges
(5) galvanized sliding bolts
8"-diameter flat-free tires
3—2 × 6 × 8" treated
10—2 × 4 × 8' treated
2—2 × 4 × 10 treated
8—2 × 2 × 8' (actual 1½"—untreated)
1—1 × 6 × 8' treated or cedar
2—T1-11 plywood siding (4 × 8 sheet)
1—½" BC plywood
6—Green PVC roof panels
Chicken wire
Door stop
Eye and ear protection
Work gloves

How to Build a Chicken Coop

Cut the six supporting posts (A, B) to length. Cut the tops at an 8° angle on your miter saw (A, B measurements are to the long edge of the angle). This will translate into a slope of 1½" for every foot.

Cut the 2 × 6 bases and then lay out the sides. Square the posts to the base and then fasten them with 2½" screws (either deck screws or self-tapping screws). The posts will be bolted to the 2 × 6s later, after the plywood sides are fastened—which will ensure that the posts are properly squared, since pressure-treated wood is often not quite straight.

Cut the front and back pieces of plywood siding. Attach the vertical edges of the plywood front (and back) to the two front and back 2 × 4 posts. The plywood should extend ½" past the 2 × 4 on both edges. Fasten the plywood with 1⅝" screws.

Screw 2 × 2s to the bottom edges of both pieces of plywood siding between the posts. Fasten it with 1⅝" screws through the face of the siding. Stand the two walls up and join them with 2 × 2s and 2½" screws. Predrill and countersink the screws, driving them at a slight angle starting at about 1¼" from the ends. This will help avoid splitting the ends. Tack on a temporary brace to help keep the walls plumb until the plywood sides are on.

Add two more 2 × 2 joists between the two walls. Cut the plywood floor, notching the corners around the 2 × 4 posts. The plywood should go right to the edge of the posts. Screw the plywood to all the 2 × 2 joists. Also screw additional 2 × 2 joists to the edges of the plywood floor to serve as nailers for the two plywood side panels.

Measure and cut the plywood siding for the angled side at the center of the coop. Attach clamps or temporary blocks to the center posts to hold the piece up as you fasten it. Align one edge of the plywood with the post and fasten it with 1⅝" screws, then align the other edge and fasten it. You may need to do a little pushing, or even shim the base if you're not working on level ground, but aligning this edge will make the coop square and rigid. Leave the other side of the coop open for now to make it easier to complete the framing.

Cut and fasten 2 × 2s to the top edges of all the plywood walls. The angle cuts for the sloping wall are 8°.

Bolt the posts to the 2 × 6 bases with 2 carriage bolts each, and then screw on the end pieces (B, C) with ⅜" × 4" exterior, self-tapping lag screws. Unless you're planning to leave the coop in one place permanently, add tires now. We used 8"-diameter flatless tires with ½" bolts for axles.

Add a 2 × 2 at the top of the open side of the coop, then screw on a 2 × 2 rafter in the center to help support the roof. Predrill or use self-tapping screws to avoid splitting the wood. Mark the location of this center rafter on the side so you can find it later, then screw on the last piece of siding.

Measure and cut ½" plywood for the roof. Fasten it to the 2 × 2s, aligning it with the edges. The plywood roof is mostly to provide additional warmth in the winter, but if you live in a warm climate, you don't have to install it since it will be covered by roofing panels.

Cut holes in the plywood siding for doors and a window. Use a jigsaw, tipping it into the plywood to get the cut started. Make the cuts as clean as possible—the cut out pieces become the doors. Cut a 20" × 36" main door, a 12" × 20" door for the chickens and a 6" × 20" window. Screw pieces of 1 × 3 around the two doors to serve as door stop, extending them about ½" into the openings.

Cut and install the three roof rafters that run from front to back, making 8° plumb cuts at the front and back. Fasten them to the 2 × 2s inside the plywood with 3" screws. Add two intermediate rafters to support the roof over the open part of the coop.

TEMPORARY SUPPORT

Finish the roof framing by screwing long 2 × 4 fascias to both sides of the structure. If you're working by yourself, screw a temporary support to one of the end rafters to hold the fascia up. After finishing the roof framing, nail 1 × 4 trim at the top of the plywood between the rafters on the front and back of the coop.

Set the doors into the openings in the siding and attach the hinges and the sliding bolts. Use shims under the doors to elevate them as you put the hinges on. **NOTE:** Some of these hinge and bolt screws may just poke through the inside of the siding. If they do, flatten them out with a file.

The ramp for the chickens combines the back of the small entry door and a longer piece of 1 × 8 that extends from the underside of the coop to the ground. The top end of the 1 × 8 ramp rests on a piece of 1 × wood fastened 2¼" down from the framing (total open area is 2¼" × 8"). When cleaning the open area or moving the coop, just slide the 1 × 8 up and under the framing. Nail small strips of wood for the chickens to hold on to every 4 or 5 inches on the 1 × 8 and the back of the door.

Measure, cut, and assemble outside corners for the coop. Use 1 × 3s for the angled side (8° cut) and 1 × 2s for the back and front sides. Nail on the corners with galvanized casing nails. Staple chicken wire over the inside of the window. In the winter, cover the window with plexiglas or clear plastic.

Assemble a door to the enclosure from 1 × 6s (top and bottom) and 1 × 4s (sides and center). Cut the door to follow the angle of the roof with 8° cuts, and glue and screw all joints. Note that the 1 × 6s are on the outside face of the door. After the glue has dried, staple chicken wire to the back. Fasten hinges near the top and bottom, then attach it to the post, leaving a ½" space at the bottom (note the ½" plywood in the photo) and ⅛" to ¼" of space on the hinge side. Nail a door stop to the handle side.

Attach chicken wire to all the sides with galvanized fence staples or self-tapping, wafer-head screws (wide, flat heads). Fasten the roof panels to the rafters using roofing screws with neoprene washers. Most roof panels come with matching foam or plastic strips that fit the corrugations—insert these along the edge of the roof to seal the gaps and help keep heat in and bugs out. Caulk any inside corners in the coop where you can see light coming through with latex caulk. Throw some hay into the coop and you're ready for the chickens.

brooder box

03

Expectant parents of both human and chicken babies get the same advice: have the nursery ready before you bring home the little one(s). You'll enjoy those precious first weeks a lot more if you're not running around like a . . . well, never mind. For chickens, a nursery is a brooder box, and it's used to provide the same essentials human newborns need: security, warmth, and nourishment. The nice thing about a chicken nursery is that you don't have worry about wall color or getting an ultrasound if you absolutely refuse to go gender-neutral (you're probably expecting all girls anyway).

A brooder box can be literally that simple—just a cardboard box, or a plastic bin, an old fish tank, even a kiddie pool. Backyard farmers often use whatever they have on hand. But a nice, sturdy wood box with a few convenience and safety features will make raising your chicks a better experience, for this brood and many more down the road. This brooder box is made with a single sheet of ¾ inches plywood and measures 36 × 36 × 17⅜ inches, enough room for housing 8 to 10 chicks up to six weeks old. Both the lid and the floor are covered with hardware cloth (wire mesh), making the box secure, well ventilated, and easy to clean.

Heat is provided by a simple clamp-on reflector light, which can be set directly atop the lid's wire mesh or clamped to the light pole at various heights for temperature control. The box lid is hinged in the back and locks in the front with a locking hasp latch. Special removable hinges make it easy to slide off the lid to get it out of the way for a thorough cleaning of the box.

A brooder box like this is a nursery to your newborn chicks, so it's essential that it be comfortable, warm, safe, and nurturing to your future chickens.

Building a Brooder Box

CUTTING LIST

Key	No.	Part	Dimension	Material
A	1	Top	¾ × 36 × 36"	¾" plywood
B	2	End	¾ × 36 × 15⅞"	¾" plywood
C	2	Side	¾ × 34½ × 15⅞"	¾" plywood
D	2	Handle	1½ × 3½ × 6"	2 × 4
E	4	Base	¾ × 1½"	1 × 2
F	1	Light mount	1½ × 3½ × 6"	2 × 4

TOOLS & MATERIALS

Circular saw
Drill
Jigsaw
Wire cutters
Staple gun
Wood glue
Sandpaper

Deck screws 1⅝", 2"
½" × ½" galvanized hardware
    cloth (36 × 70" min.)
Heavy-duty ½" staples
½"-diameter × 36"-long
    hardwood dowel
(2) Separable lid hinges with screws

Locking hasp latch with screws
Clamp-on light fixture
Zip ties (optional)
Eye and ear protection
Work gloves

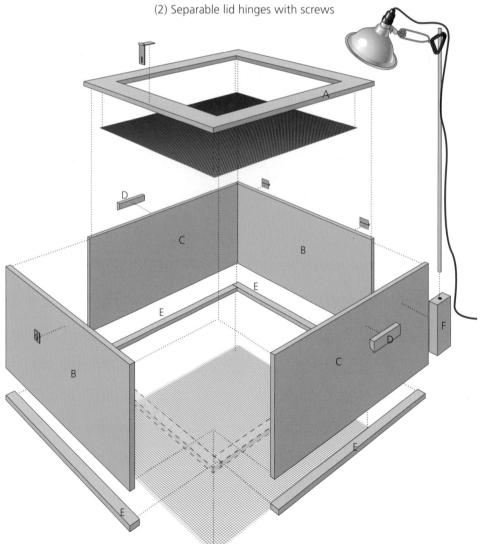

How to Build a Brooder Box

Make the lid cutout by marking four lines 4" from the outside edges of the lid piece. Carefully lower your circular saw onto one end of the line and cut to the other end. Finish the cuts at the corners with a jigsaw. **TIP:** Roughly lay out the parts on the plywood sheet before marking and cutting the pieces one at a time.

Create the two handles, beveling one long edge at 15°. Cut the strip in half to end up with two 6"-long handles. Install the handles with glue and 1⅝" deck screws so the top edges of the handles are about 2" below the top of the box sides. The long beveled edges face down and toward the box side, creating an easy-to-grab lip for each handle. **Inset:** Construct the light pole mount by drilling a ½" centered hole in the end of a scrap 2 × 4. Ream the hole until the dowel fits easily and can be slipped in and out to facilitate box cleanings.

Assemble the box frame using glue and 2" deck screws. Fit the front and back pieces over the ends of the sides, drill pilot holes, then apply glue and fasten each joint with five evenly spaced screws. Use a framing square to make sure the corners are square as you work. Sand all exposed edges of the box and lid.

Use wire cutters to cut the box bottom and lid cover from ½" hardware cloth. Cut the bottom to 35¾ × 35¾"; cut the lid cover 34 × 34". Screw 1 × 2 cleats to the bottom edges of the plywood box frame, then staple on the bottom cloth with heavy-duty ½" staples. Keep the edges of the mesh ⅛" from the outside edges of the box on all sides. Staple the lid cover to the underside of the lid frame, keeping the mesh 1" from all outside edges. Use plenty of staples on the cover—and staple close to the edges of the cutout—to prevent predators from pushing down the mesh.

Install the box lid using two separable hinges (also called removable or lift-away hinges). Depending on the hinge design, you may want to recess the hinge plates on the box by cutting shallow mortises, using a wood chisel. Add a locking hasp latch to the front side so the lid can lock securely to the box front. Clamp the light fixture to the dowel and plug it in to test the setup.

large farm animals

Raising farm animals can be a challenge and it is a serious responsibility, but once you've taken the plunge, the rewards can be so great that it is very difficult to go back. But it's important to select animals that will be comfortable on your property.

How hot or cold is the region where you live? While you can make accommodations by choosing the right shelter for your animal, some extreme climates simply are not suitable for every animal. Before you purchase livestock, contact local veterinarians and find out whether they will treat the species you plan to buy. If you cannot find a vet in your area who will treat the llama you're planning to purchase, find out how far you'll have to drive to reach someone who will, and weigh this into your decision. Also, think through the logistics of how you will transport your animal and the time it will take to reach the vet.

Lastly, check with your municipality to find out what the limitations are when it comes to owning livestock. If you live in the city, you may need to request permission from your neighbors to house animals, and some species may not be allowed at all. There may also be setback requirements that apply to animal housing—take these into consideration when planning out the site for your animals' home.

A barn is not necessary for every species, though most animals require some type of shelter for protection from the elements and, perhaps, for sleeping. The type of roof you put over animals' heads depends largely on where you live. In the chilly north, animals will require a more sturdy, draft-resistant abode than in the hot and dry southwest. Build your shelter to suit your climate—in cold climates, your shelter should be designed to keep animals warm, and in warm climates, designed to keep animals cool. Tailor your shelter to meet your animals' needs. For example, pigs require separate space for eating and sleeping, whereas sheep and goats just need a dry shelter to protect them from the elements.

Owning big animals can be rewarding, but they can require lots of work, and you'll need strong fencing to keep them from wandering away and exploring the neighborhood.

Pigs love attention and are infinitely curious. Be sure to build a strong fence around your pigpen to keep your pigs contained.

○ Pigs

Pigs are personable and intelligent. They also offer a well-rounded learning experience for new animal owners: lessons on the importance of feed mix, daily pigpen cleaning to prevent disease (and smell!), and good old-fashioned recycling. Their composted byproduct is rich fertilizer for your garden.

Pigs don't require a great deal of land, but they do need a dry, draft-free shelter to protect them from the elements. Be sure to prepare this space before bringing home your new pets. You can dedicate a portion of an existing building, such as a barn or shelter, or construct a simple outbuilding. Ideally, the floor should be concrete and sloped for optimal drainage during daily hose-outs. A dirt floor is also fine as long as you replace hay bedding daily.

Pigs also require a separate sleeping and feeding area. A 5 foot by 5 foot square sleeping area will accommodate two pigs. The feed area should be twice this size and contain the feed trough, a watering system, and a hose connection. This serves the double purpose of instant water refills for thirsty pigs and accessibility to your number one pigpen-cleaning tool.

Contain your pigs with a secure fence of woven wire or permanent board. The fence should be about 3 feet tall. Pigs are immensely social animals and love the spotlight, so be sure to spend time with them to help them grow healthy and happy.

Goats are high-spirited, lovable, and mischievous animals that love to play. Watch out, though—your goats will try to outsmart you (or your fence!). Goats are a great source of milk, wool, and meat.

○ Goats

Goats are mischievous class clowns with boundless energy and a lovable nature. They are a great source of both milk and meat, and do not require a vast amount of resources, food, or shelter. They are ruminants that enjoy munching twigs and leafy brush—but if you will be raising goats for their milk, feed them with a forage of hay and grain to preserve their milk's taste. Some goat breeds are also a great source of fibers for fabrics, such as mohair and angora. You'll want to watch these breeds' diets carefully, as their coat will be affected by their diet. Feed them only quality pasture or hay, along with plenty of fresh, clean water.

There's no need to build a fancy house for goats, as they fare well in pretty much any dry, draft-free quarters. Goats are prone to respiratory problems triggered by a moist environment, so avoid heating that can result in condensation. House goats in a three-sided barn, shed, or a shared barn with other animals. It's a good idea to invest your savings on shelter in quality fencing, however. Woven-wire pasture fencing is ideal, and additional strands of barbed or electrical wire will discourage curious goats from escaping.

○ Sheep

Sheep are affectionate animals raised most often for their high-quality wool that can be spun into yarn and made into warm textiles. Choose a breed of sheep with a wool density that correlates with the weather where you live—in the north, choose sheep with extra "lining" in their coats; in temperate or arid climates, sheep with fine wool and hair prosper.

When purchasing sheep, look for healthy feet. From the front view, legs and hooves should align, as opposed to being knock-kneed, splayfooted, or pigeon-toed. Check the animal's bite, and be sure there are no udder lumps or skin lesions. Sheep do need to be shorn every spring before the weather heats up, so take good care of your sheep's wool and sell or spin it after it is collected. To care for your sheep (and their wool), make sure they have good nutrition, well-managed pastures, and vaccinations. Sheep need a sturdy fence and some type of shelter, though existing buildings on your land will suit them just fine. They are easy targets for predators, such as coyotes, so make certain your fence is secure.

Sheep travel in close-knit packs, are spooked easily, and are an easy target for predators, so make sure your sheep are protected by a secure fence.

○ Alpacas

Alpacas are smaller cousins to llamas and camels and are an approachable, friendly species, which makes them appealing to landowners who want to begin caring for animals. They won't challenge your fencing or trample your pasture. They also require little feed—about a third less than a sheep. Alpacas grow thick coats that are five times warmer than wool and far more durable. Yarn spinners covet alpaca fiber, homeowners admire these loving pets, and investors appreciate the potential returns these valuable creatures promise.

Fencing you'll build for alpacas is designed more to keep predators out than to keep alpacas in. These animals are not ambitious escape artists—not nearly as tricky as goats. But predators can represent a threat to sensitive alpacas, so it's a good idea to install strong perimeter fencing that is at least 5 feet tall. Separate females and males with fencing. Females and their newborns must have separate quarters from the rest of the pack, but do not completely isolate them from the group. A three-sided shelter is adequate for alpacas, which are accustomed to rugged, cold climates. Heat is more of a concern for these animals, and their insulating fiber coats are no help in keeping them cool in summer. A misting system or fans in the alpaca shelter will prevent them from overheating.

Lock eyes with a quizzical alpaca and you'll feel like you are being probed for information. Alpacas are gentle animals that are easy to care for and produce soft, extremely warm coats that can be shorn and sold to textile makers.

top-bar beehive

05

TIP **5 Ways to Keep Your Bees Safe & Healthy**

1. Avoid using insecticides in your garden—Many are long-lasting and toxic to bees.

2. Buy seeds that are not treated with insecticides—Some coated seeds may cause the entire plant to become toxic to bees. Check seed packets carefully.

3. Mix your own potting soil and compost—Some composts and potting mixes sold at garden centers contain insecticide that is highly toxic to bees and other insects, and will eventually pollute all of your soil. Make your own compost and mix with natural additives for potting plants.

4. Plant bee-friendly flowers—Buy wildflower seed mix and plant in uncultivated areas to create small sections of wild, natural habitat for your bees.

5. Provide a home for bees—Whether you're a blossoming beekeeper or not, it's easy to provide a home for bees! Provide a simple box as a place for feral bees to nest, or start your own hive.

Backyard beekeeping makes more sense today than ever before. Not only are honey bees necessary for pollinating plants and ensuring a better fruit set and bigger crops, they produce delicious honey and valuable beeswax. And recently, the world bee population has experienced a mysterious and concerning dropoff in numbers. Getting homeowners to cultivate a bee colony is a helpful component of the preservation strategy.

In many ways, tending bees is like growing food. There is an initial flurry of activity in spring, followed by ongoing maintenance in the summer, and then harvest in the fall. There is prep work you'll need to complete before you begin and there is a learning curve—you'll need to spend more time with your bees in the beginning until you learn how it's done. Beekeeping is not necessarily an expensive hobby, but with higher-end operations, purchasing the hive, some blue ribbon bees, and all the necessary equipment can require a significant financial investment.

Keeping bees will help you have a better garden, more fruits and vegetables, and honey in the kitchen—even beeswax candles, skin creams, and other natural cosmetics. And, by building a top-bar beehive, you're creating a safe home and enabling one of our earth's most necessary and miraculous species to thrive.

Honey and beeswax are the two commodities a functioning backyard beehive will yield. If your primary interest is honey, build a traditional stacking-box style beehive (see page 50) in which the bees expend most of their energy filling the premade combs with sweet honey. If it's beeswax you seek, make a top-bar hive like the one shown on the following pages.

Top-Bar Hive

Expert beekeeper Phil Chandler insists that beekeeping should be a very simple pursuit, largely because the bees do almost all of the work for you. Chandler, who maintains a website called The Barefoot Beekeeper (see Resources, page 348), is an advocate for natural beekeeping and has designed a top-bar hive that you can build yourself using simple materials. This hive (see pages 47 to 49) is designed to enable the bees to build their own comb, instead of relying on a premade comb.

The top-bar hive is simple in its construction and, unlike the traditional stacked box Langstroth hive, does not require that you lift heavy boxes to check on your hive's progress, which disturbs the bees within. Rather, you can simply remove the hive roof and inspect the bars one by one without disturbing the rest of the hive. Storage is minimal for a top-bar hive, as there are no supers needed. And, it is not necessary with this hive design to isolate the queen.

This simple top-bar beehive design is a warm and safe home for bees that is easily adjustable to accommodate a growing hive. This design also greatly simplifies the inspection process and minimizes the amount of equipment needed to keep and maintain bees.

Building a Top-Bar Beehive

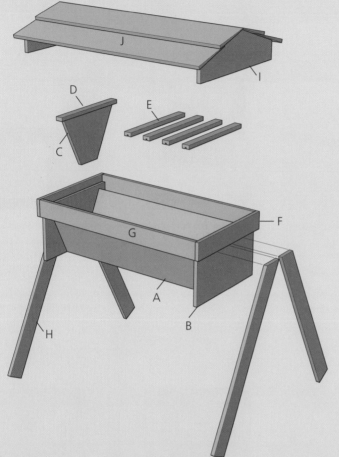

TOOLS & MATERIALS

Lumber (1 × 2, 1 × 3, 2 × 4, 1 × 12)

Carpenter's square

Pencil

Circular saw or table saw

Socket wrench

Exterior-grade construction adhesive

Caulk gun

Clamps

Drill

Tape measure

Hammer

Handsaw

1¼", 2", 2½" deck screws

Stainless-steel or plastic mesh

Roofing nails or narrow crown staples

Eye protection

1" holesaw

⅜ × 2" galv. lag bolts with washers and nuts

Eye and ear protection

Work gloves

CUTTING LIST

Key	Part	No.	Dimension	Material
A	Side panel	2	¾ × 11¼ × 36"	Cedar
B	End panel	2	¾ × 11¼ × 19"	Cedar
C	Insert	2	¾ × 11¼ × 15"	Cedar
D	Insert cap	2	¾ × 1½ × 17"	Cedar
E	Top Bar	20	¾ × 1½ × 17"	Cedar
F	Frame end	2	¾ × 3½ × 21"	Cedar
G	Frame side	2	¾ × 3½ × 36"	Cedar
H	Leg	4	¾ × 1½ × 36"	Cedar
I	Cap end	2	¾ × 7¼ × 23"	Cedar
J	Roofing	4	⅝ × 5⅞ × 40"	Cedar bevel lap siding

How to Build a Top-Bar Beehive

Lay out cutting lines for the insert panels on a piece of 1 × 12 cedar stock. The trapezoid-shaped panels (sometimes called followers) are meant to slide back and forth within the hive cavity, much like a file folder divider. This allows the beekeeper to subdivide the hive space as the honeycombs accumulate. The shape should be 15" wide along the top and 5" wide along the bottom (See diagram, page 47).

Cut the insert panels to size and shape and then attach a top cap to the top edge of each panel. The 1 × 2 caps, installed with the flat surface down, should overhang the panels by 1" at each end. Use exterior-rated wood glue and 2" deck screws driven through pilot holes to attach the tops. Also cut 20 top bars from the same 1 × 2s. Use a router or table saw to cut a ¼ × ¼" groove in the bottom of each top bar (inset). The bees use these grooves to create purchase for their hanging honeycombs.

Secure the two insert panels upside down on a flat work surface and use them to register the side panels so you can trace the panel locations onto the end panels. Center the end panels against the ends of the side panels, making sure the overhang is equal on each side. Outline the side panel locations, remove the end panels, and drill pilot holes in the outlined area.

Attach the end panels to the side panels with glue and 2½" deck screws driven through the pilot holes in the end panels.

Cut the parts for the frame that fits around the top of the hive box and fasten them with glue and 1¼" deck screws. The top of the frame should be slightly more than ¾" above the tops of the side panels to provide clearance for the top bars, which will rest on the side panel edges.

Attach the legs. First, cut 36"-long legs from 1 × 4 stock and place them over the box ends as shown in the diagram on page 47. Mark cutting lines where the leg tops intersect with the bottom of the frame. If your hive will be on grass or dirt, leave the bottom ends uncut to create a point that will help stabilize the hive. If your hive will be on a hard surface, cut the ends so they are parallel to the tops and will rest flush on the ground. Attach the legs with two or three 3⁄8 × 2" galvanized lag bolts fitted with washers and nuts.

Drill entrance holes and attach the box bottom. On one side panel, drill three 1"-dia. bee entrance holes 2" up from the bottom of the hive. One hole should be centered end to end and the others located 3" away from the center. On the other side, drill a 1"-dia. hole 2" up from the bottom of the hive and 5" from each end. Attach a steel or plastic mesh bottom with roofing nails or narrow crown staples.

Make and install the lid. You can design just about any type of covering you like. Here, a frame with a gable peak is made from cedar stock and then capped with beveled-lap siding (also cedar). The overlap area where the siding fits along the peak ridge should be sealed with clear exterior caulk. Add the inserts and top bars and then fit the lid frame around the box top frame.

traditional beehive

06

A traditional beehive is an ideal addition to the self-sufficient homestead. Create a welcoming home for a colony of bees and you ensure pollination for the plants in your garden, add to the health of your local ecosystem, and produce a bumper crop of beekeeper's gold—better known as sweet, delicious honey.

The 10-frame hive design shown here is familiar to anyone who has passed an active orchard or seen beekeepers in action. It is formally known as a Langstroth Hive, and is the standard for beehives. It features a basic construction meant to accommodate removable frames into which bees build honeycombs. Depending on where the frame is in the hive, the bees will use the honeycomb for brooding—to raise new bees—or to hold honey. The discovery that led the Reverend Lorenzo Lorraine Langstroth to develop this design was that a specific spacing—5/32-inches to be exact—was ideal to allow bees to move around in the hive. The space is too small for bees to fill in with honeycomb, and too large for them to try to close it up with the wax-like material propolis. That meant that the frames were left free to be removed. The beekeeper could easily take out a diseased section or remove frames to harvest honey. The design has been used around the world ever since.

Although it may look like a complex project, a beehive is fairly simple in construction. Building a beehive begins at the bottom, with what is know as the "bottom board." There are three types of bottom boards: solid, screened, and slatted. A screened bottom board is used where varroa mites are active and where the bees will pick them up. The screen allows for the mites to fall down and through the screen, so that they can't get back to the bees. The slatted bottom rack is a complex construction that further protects the hive by improving air circulation, adds additional cluster space for the bees, and prevents swarming. Any bottom board is built slightly longer than the hive bodies, to leave room for bees to enter and leave. This project describes how to build a basic solid bottom board. Once you've built that, it's easy to find plans and construct a screened bottom board or slatted bottom rack, should you think either of those necessary for your hive. Some beekeepers use a combination, especially in bigger and busier hives.

A Simple, Easy-to-Build Hive Stand

You'll need to use some type of stand to get your hive up off the ground and away from potential predators. The stand should also maintain a bit of air circulation around the hive. This basic unit shown here can be made in minutes with a few pieces of lumber and some screws. Mark and drill pilot holes through the brace faces down into the base 4 × 4s. Screw the braces to the base using 4" deck screws and you're finished! You can add 4 × 4 legs to the base if you prefer to the keep the hive even higher.

A traditional beehive such as this can serve as the perfect home for your own busy colony . . . and a gathering place for delicious honey.

51

Sitting atop the bottom board are two or more body sections. There are actually two types of these: the deeper "hive bodies" that house the brood, or young bees (in all stages), and the upper honey "super," which contains the honeycombs filled with honey. Both house rows of suspended frames.

The boxes are simple in construction but can be built in a number of different ways. We've chosen the simplest style of construction—butt joints—but if you have woodworking experience and want to take your hive construction to the next level, you can use rabbeted or dovetailed joints. Both will be stronger than butt joints. The number of hive bodies and supers you include depends on a couple of factors. A regular or small population of bees will normally require only one hive body. If you add bees, you'll need another. Supers range in depths. If

you've chosen thin supers to make the honey frames easier to handle, or if you have enough bees to require two bodies, you'll want more than one super. In any case, we've included one hive body and one super in this project for purposes of clarity in illustrating the details.

No matter how many bodies or supers you include, they'll need to be capped by a top. Most hives include an inner cover with a hole that permits movement in and out of the cover, and an outer cover that fits completely over the top structure to seal the hive and protect it from predators and the elements. You can also choose a screened inner cover that allows for better air circulation and enables you to inspect your hive without unduly disturbing the bees or having to remove the inner cover. Lastly, we've included instructions for a very basic hive stand.

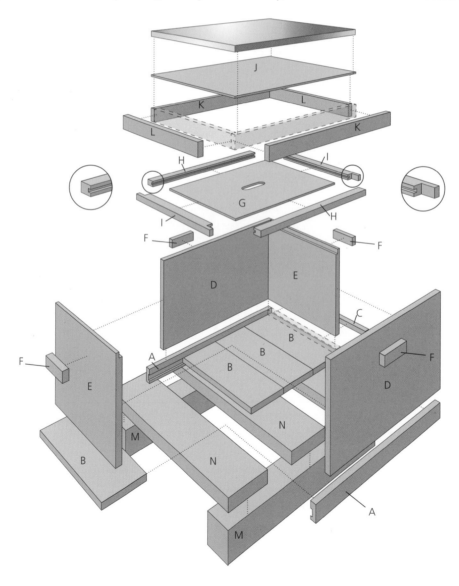

Building a Beehive

Hives are not placed on the ground for a variety of reasons, and a stand like the one shown here ensures that the hive is portable, should it need to be moved. You can choose a more fixed design by attaching legs to the bottom of the bottom board, or simply place your hive on a table or sawhorses.

You can make your own beehive frames from scratch, but given the intricate and repetitive woodworking involved, the vast majority of hive owners buy premade frame kits for the standard frame, called "Dadant frames" (available online or from stores that stock beehive equipment). The kits include all the components necessary to build the frames, along with detailed instructions.

CUTTING LIST

Bottom Board

Key	No.	Part	Dimension	Material
A	2	Side	1⅞ × ¾ × 22"	1 × 3
B	4	Bottom	¾ × 5½ × 15¼"	1 × 6
C	1	Filler strip	¾ × ¾ × 14¾"	1 × 1

Hive Body

Key	No.	Part	Dimension	Material
D	2	Side	¾ × 11¼ × 19⅞"	1 × 12
E	2	End	¾ × 11¼ × 14¾"	1 × 12
F	4	Handle	¾ × 1½ × 4"	1 × 2

Inner Cover

Key	No.	Part	Dimension	Material
G	1	Center board	¼ × 14¼ × 17⅞"	¼" plywood
H	2	Side trim	¾ × 1¼ × 18⅞"	1 × 2
I	2	End trim	¾ × 1¼ × 16¼"	1 × 2

Outer Cover

Key	No.	Part	Dimension	Material
J	1	Top	¼ × 21⅞ × 18¼"	¼" plywood
K	2	Side	¾ × 1¾ × 21⅞"	1 × 3
L	2	End	¾ × 1¾ × 16¾"	1 × 3

Hive Stand

Key	No.	Part	Dimension	Material
M	2	Base	3½ × 3½ × 24"	4 × 4
N	3	Brace	1¾ × 5½ × 24"	2 × 6

TOOLS & MATERIALS

Cordless drill and bits
Miter saw
Tablesaw and dado blade
Jigsaw
23⅜ × 20" aluminum flashing
  or a premade hive top
Deck screws 1¼", 1⅝", 2", 4"
¾" sheet metal screws or roofing nails
Waterproof wood glue
Sandpaper
Bar or pipe clamps
Aviation snips
Rubber mallet
Straightedge
Eye and ear protection
Work gloves

How to Build a Beehive

On a table saw, cut the dado in each side piece using a dado blade. (You can also use a router with a ¾" straight bit.) Start with a 1 × 6 to make it easier and safer to cut the dados, then rip the board to 1⅞" wide. Set the fence ¾" from the blade, with the blade height ¼". For safety, clamp a temporary top of scrap plywood to the table and raise the dado blade through it. Make the dado cut slightly wider than ¾". It's a good idea to test cut a scrap piece and check it for fit with the bottom boards. Then rip the dadoed sides to 1⅞" wide.

Optional: A tongue-and-groove construction will be stronger than the butt joints used in this project, but it is also a lot more work to cut all pieces with the tongues and grooves. A compromise that improves the strength of the bottom board but takes less time is lap-joint construction. Set a table saw blade height to ⅜" and set the fence to ⅜" measured to the outside of the blade. Cut top and bottom half laps on each side of the middle boards, and one top and one bottom lap on either end board.

Dry fit the pieces to ensure there are no gaps. Apply the glue along the inside edges of the boards and in the dado of the side pieces. Assemble the pieces and wipe away any glue squeeze out. Clamp the entire assembly using bar or pipe clamps, until the glue dries.

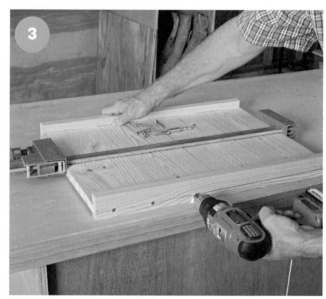

Use a square or straightedge to mark the center of the dado along the outside face of the side pieces. Drill pilot holes through the outside faces of the side pieces into the edges of the bottom boards. Drill two holes per side for each outside board and one hole for the inside boards, then drive 4" screws through the side pieces into the bottom boards. Do the same with the filler strip at one end of the construction.

Set up the table saw to cut rabbets along the top edge of both end pieces. Set the fence to ⅜" measured to the outside of the blade, and the blade height to ⅝". Push the end piece through on end. (A featherboard, shown here, helps guide the workpiece along the fence.) To finish the cut, change the fence to ⅝" to the outside of the blade and the blade height to ⅜", and push the piece through lying flat. Test cut a scrap piece, then cut the rabbets along the top edge of both end pieces.

Rip a 15° miter along the edge of a 2-ft. long 1 × 4 or 1 × 6, then cut it into 1½" × 5" handles. Measure and mark the handle location centered across the width of the ends and sides, 2" down from the tops. Hold the handle in position on each piece, drill two pilot holes for each handle, then screw the handles on with 1¼" deck screws.

Apply exterior glue down the inside edges of each side piece. Assemble the box and clamp it into position using bar clamps (a helper makes this process much easier). Check for square and adjust the box if necessary while the glue is still wet.

Drill pilot holes along the face of the side pieces, where they overlap the edges of the ends. Drive a few 2" screws near the top to hold the sides together, then remove the clamps and add the rest of the screws, spacing them about every 2".

(continued)

How to Build a Beehive (continued)

Rip the side and end trim pieces down to 1¼" wide. You can cut all the pieces from one 72" 1 × 2. Cut a ¼" dado at the center along one edge of each trim piece (set the fence at ¼" from the blade). Use a dado blade or just make a few passes with a regular blade. The dado should be ⁵⁄₁₆" deep.

Cut the rabbets on the ends of the end pieces using the table saw. The rabbets are cut 1¼" deep by ¾" in wide. Use a simple sliding jig made from a 2 × 6 and a clamp to complete the 1¼" cut. Then set the two boards on edge and make the ¾" cut using the sliding miter gauge. Make sure to move the fence away from the board so that the cutoff pieces don't bind and jam against the blade.

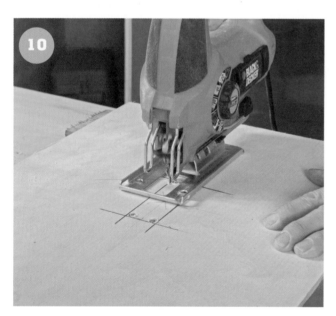

Measure and mark centerlines on the plywood cover. Mark a 4 × 1½" opening at the center of the board. Drill starter holes at the corners and cut out this hole using a jigsaw. Smooth the edges with sandpaper.

Dry fit all the pieces together and make adjustments as necessary. Lay a bead of glue in the dadoes and assemble the cover, clamping it tightly with bar clamps. When the glue has dried, drill pilot holes through the outside edge of the rabbets on the end pieces, into the ends of the side pieces. Screw them together with 1⅝" deck screws.

Drill pilot holes through the side faces into the ends of the end pieces for the butt joints. Screw the top frame together using two 1⅝" deck screws per joint. Check for square as you work to ensure the frame will exactly align with the top.

Align the top to the frame on all sides, then screw the top down with predrilled 1¼" deck screws on all sides.

Use the frame upside down to mark the bend lines on the sheet metal top, then cut out the 45° corners with aviation snips. Place the metal over the top and clamp it firmly in place at the bend line with a straight 2 × 4. Bend the edge over with a piece of wood, then use a rubber mallet to make the bend sharper.

Screw or nail the metal top to the wood sides with ¾" galvanized fasteners spaced every few inches.

nourishing your garden

In many ways, the self-sufficient home revolves around the garden. Growing and harvesting your own produce enables you to declare some measure of independence from the corporate food chain. It also allows you to control what goes into what you grow, meaning that you'll eat less pesticides, chemicals from processed fertilizer, and other contaminants. And don't forget the money you'll save by raising your own organic produce.

But self-sufficient gardening is something more as well. It's the chance to get involved with your food and truly "get back to the land." There is nothing quite like a few sore muscles and the knowledge that at the end of a day of gardening, you've done simple, good work. You soak up sunshine, get the most wholesome form of exercise you can get, and achieve something tangible and positive. You could hardly ask for a more rewarding outdoor activity.

Of course, gardening involves much more than simply digging in the dirt and dropping a seed in a hole. In fact, it starts with that dirt. Healthy soil will be the foundation of all you do in the garden, and without it, your labor is likely to be much less productive. Much of the fertilizer sold at home centers is like junk food for plants—a quick, sugary rush that leaves your soil more depleted afterward. The best way to build up your soil and make it more productive is with compost, which provides rich, long-lasting nourishment for your soil, which in turn nourishes your plants in the best way possible. You can buy bags of compost, but you can also make your own by recycling your leaves, grass clippings, kitchen waste, coffee grounds, chicken bedding, and more. Just mix them up a compost bin and in a month you'll have rich, nutritious soil to spread around the garden.

Rainwater is another great resource for the garden and lawn, but too many homeowners let rainwater drain away or accumulate where it's not needed (such as around the foundation), and end up relying on their tap water to keep plants green. In this section we'll show you some easy ways to capture your rainwater and use it, instead of wasting it or just letting it run down the storm sewer. You'll also learn about sub-irrigated planters, which are the best way to deliver a steady supply of water to your plants if you're using containers or raised beds.

compost bin

The byproducts of yard maintenance and food preparation accumulate rapidly. Everyday yardcare alone creates great heaps of grass clippings, trimmings, leaves, branches, and weeds. Add to this the potato peelings, coffee grounds, apple cores, and a host of organic kitchen leavings. The result is a large mass of organic matter that is far too valuable a resource to be simply dumped into the solid waste stream via curbside garbage collection. Yard waste and kitchen scraps can be recycled into compost and incorporated back into planters or garden beds as a nutrient-rich (and cost-free) soil amendment.

Compost is nature's own potting soil, and it effectively increases soil porosity, improves fertility, and stimulates healthy root development. Besides, making your own soil amendment through composting is much less expensive than buying commercial materials.

So how does garbage turn into plant food? The process works like this: organisms such as bacteria, fungi, worms, and insects convert compost materials into humus, a loamy, nutrient-rich soil. Humus is the end goal of composting. Its production can take a couple of years if left undisturbed, or it can be sped up with some help from your pitchfork and a little livestock manure.

Although composting occurs throughout nature anywhere some organic matter hits the earth, in our yards and gardens it is always a good idea to contain the activity in a designated area, like a compost bin. Functionally, there are two basic kinds of bins: multi-compartment compost factories that require a fair amount of attention from you and hidden heaps where organic matter is discarded to rot at its own pace. Both approaches are valid and both will produce usable compost. The compost bin project

Composting turns yard waste and kitchen scraps into a valuable soil amendment.

shown on page 64 is an example of the more passive style. At roughly 1 cubic yard in volume, it can handle most of your household organic waste and some garden waste. If you have a higher volume of organic waste, you may want to use a two- or three-bin approach (see page 68), which allows you to have piles in different stages of decomposition.

Compost Variables

Air: The best microbes for decomposing plant materials are aerobic, meaning they need oxygen. Without air, aerobic microbes die and their anaerobic cousins take over. Anaerobic microbes thrive without oxygen and decompose materials by putrefaction, which is smelly and slow. Your goal is aerobic activity, which smells musty and loamy, like wet leaves. Improve air circulation in your compost bin by ensuring air passageways are never blocked. Intersperse layers of heavier ingredients (grass clippings, wet leaves) with lighter materials like straw, and turn the pile periodically with a garden fork or pitchfork to promote air circulation.

Water: Compost should be as wet as a wrung-out sponge. A pile that's too wet chokes out necessary air. A too-dry pile will compost too slowly. When adding water to a compost pile, wet in layers, first spraying the pile with a hose, then adding a layer of materials.

Temperature: A fast-composting pile produces quite a bit of heat. On a cool morning, you might notice steam rising from the pile. This is a good sign. Track the temperature of your pile and you'll know how well it's progressing. Aim for a constant temperature between 140 and 150 degrees Fahrenheit, not to exceed 160 degrees. To warm up a cool pile, agitate it to increase air circulation and add nitrogen-dense materials like kitchen waste or grass clippings. A pile about 3 feet high and wide will insulate the middle of the pile and prevent heat from escaping. You'll know the compost process is complete when the pile looks like dirt and is no longer generating extraordinary heat.

Called "black gold" by home gardeners, compost can be generated on-site and added to any planting bed, lawn, or container for a multitude of benefits. Sifting the compost before you introduce it to your yard or garden is recommended.

What to Compost, What Not to Compost

Vegetable plants soak up the materials that make up your compost, and these materials will play a vital role in the development of the vegetables that will grace your dinner table! When in doubt as to what should or shouldn't go into your compost pile for your garden, follow these general guidelines.

Great Garden Compost	Not for Compost, Please
"Clean" food scraps—including crushed eggshells, corncobs, vegetable scraps, oatmeal, stale bread, etc.	Fatty or greasy food scraps—including meat waste, bones, grease, dairy products, cooking oils, dressings, sandwich spreads, etc.
Vegetable and fruit peelings and leftovers	Fruit pits and seeds—these don't break down well and can attract rodents.
Coffee grounds and filters, tea leaves and tea bags	Metal. Remove the tea bag staples before composting!
Old potting soil	Diseased plant material
Lawn clippings	Weeds—these will only sprout in your garden! Kill the weed seeds and salvage the compostable bits by baking or microwaving the plants before adding them to your compost bin.
Prunings from your yard, chopped up in small pieces	Big chunks of yard debris or plants that are diseased or full of insect pests
Shredded leaves and pine needles	Any plant debris that has been treated with weed killer or pesticides
Shredded newspaper and telephone books—black and white pages only	Glossy color ads or wax-coated book covers
White or brown paper towels and napkins	Colored paper towels and napkins
Wood ash—use sparingly	Coal ash
Cardboard	Pizza boxes or other wax-coated food boxes
Livestock manure	Cat, dog, or other pet waste, which may contain meat products or parasites
Sawdust, wood chips, and woody brush	Sawdust from wood treated with preservatives
Straw or hay—the greener, the better!	
Wilted floral bouquets	

Building a Compost Bin

CUTTING LIST

Key	Part	No.	Dim.	Material
A	Post	8	1½ × 1¾ × 48"	Cedar
B	Door rail	2	1½ × 3½ × 16"	"
C	Door rail	2	1½ × 1¾ × 16"	"
D	Door stile	4	1½ × 1¾ × 30½"	"
E	Panel rail	3	1½ × 3½ × 32½"	"
F	Panel rail	3	1½ × 1¾ × 32½"	"
G	Panel stile	3	1½ × 3½ × 30½"	"
H	Infill	16	¾ × 1½ × 30½"	"
I	Filler	80	¾ × 1½ × 4"	"
J	Panel grid	12	¾ × 1½" × Cut to fit	"

Key	Part	No.	Dim.	Material
K	Grid frame-v	16	¾ × 1½" × Cut to fit	Cedar
L	Door frame-h	4	¾ × 1½" × Cut to fit	"
M	Top rail-side	2	1½ × 1¾ × 39"	"
N	Top rail-back	1	1½ × 1¾ × 32½"	"
O	Front spreader	1	1½ × 3½ × 32½"	"

TOOLS & MATERIALS

½" galvanized hardware cloth 36" by 12'
U-nails (fence staples)
2 pairs 2 × 2" butt hinges
2½", 3" deck screws
Pipe or bar crimps
Exterior wood glue
Galvanized finish nails
Exterior wood sealant
Table saw or circular saw
Eye and ear protection
Work gloves

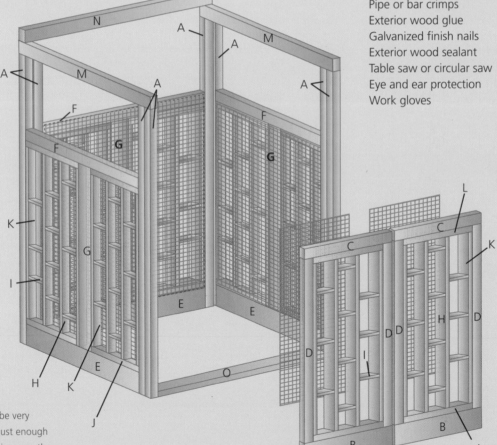

A compost bin can be very plain, or it can have just enough decorative appeal to improve the appearance of a utility area.

How to Build a Compost Bin

Prepare the wood stock. At most building centers and lumber yards, you can buy cedar sanded on all four sides, or with one face left rough. The dimensions in this project are sanded on all four sides. Prepare the wood by ripping some of the stock into 1¾"-wide strips (do this by ripping 2 × 4s down the middle on a tablesaw or with a circular saw and cutting guide).

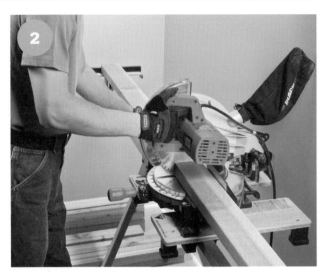

Cut the parts to length with a power miter saw or a circular saw. For uniform results, set up a stop block and cut all similar parts at once.

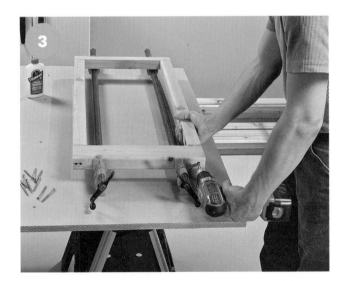

Assemble the door frames. Apply exterior-rated wood glue to the mating parts and clamp them together with pipe or bar clamps. Reinforce the top joints with 3" countersunk deck screws (two per joint). Reinforce the bottom joints by drilling a pair of ⅜"-dia. × 1"-deep clearance holes up through the bottom edges of the bottom rails and then driving 3" deck screws through these holes up into the stiles.

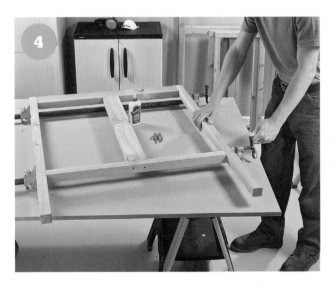

Assemble the side and back panels. Clamp and glue the posts and rails for each frame, making sure the joints are square. Then, reinforce the joints with countersunk 3" deck screws.

(continued)

How to Build a Compost Bin (continued)

Hang the door frames. With the posts cut to length and oriented correctly, attach a door frame to each post with a pair of galvanized butt hinges. The bottoms of the door frames should be slightly higher than the bottoms of the posts. Temporarily tack a 1 × 4 brace across both door bottom rails to keep the doors from swinging during construction.

Join the panels and the door assembly by gluing and clamping the parts together and then driving 2½" countersunk deck screws to reinforce the joints. To stabilize the assembly, fasten the 2 × 4 front spreader between the front, bottom edges of the side panels. Make sure the spreader will not interfere with door operation.

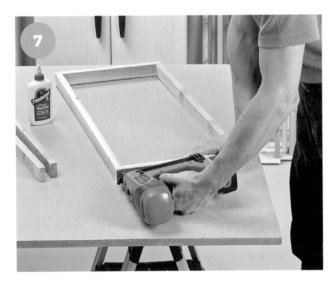

Make the grids for the panel infill areas. Use 1 × 2 cedar to make all parts. Use exterior glue and galvanized finish nails to connect the horizontal filler strips to the vertical infill pieces. Vary the heights and spacing of the filler for visual interest and to make the ends accessible for nailing.

Frame the grids with 1 × 2 strips cut to the correct length so each frame fits neatly inside a panel or door opening. Install the grid frames in the openings, making sure all front edges are flush.

Attach the top rails that conceal the post tops and help tie the panels together. Attach the sides first using exterior glue and galvanized finish nails. Then, install the back rail on top of the side rails. Leave the front of the project open on top so you can load, unload, and turn over compost more easily.

Line the interior surfaces of the compost bin with ½" galvanized hardware cloth. Cut the hardware cloth to fit and fasten it with fence staples, galvanized U-nails, or narrow-crown pneumatic staples (⅝" minimum) driven every 6" or so. Make sure you don't leave any sharp edges protruding. Grind them down with a rotary tool or a file.

Set up the bin in your location. It should not be in direct contact with any structure. If you wish, apply a coat of exterior wood sealant to all wood surfaces—use a product that contains a UV inhibitor. **TIP:** Before setting up your compost bin, dig a hole just inside the area where the bin will be placed. This will increase you bin's capacity.

A fast-burning compost pile requires a healthy balance of "browns" and "greens." Browns are high in carbon, which food energy microorganisms depend on to decompose the pile. Greens are high in nitrogen, which is a protein source for the multiplying microbes.

two-bin composter

When it's time to get serious about composting, a multiple-bin system is the way to go. They're designed to produce a large volume of compost in a short time. The idea is to develop a nice, big heap in one bin, then start turning it over by shoveling it into the neighboring bin, then back to the first bin, and so on. Turning greatly speeds decomposition (plus, it gives you a little exercise in the process). Depending on the compost materials, turning is recommended every 5 to 10 days. A two-bin composter lets you flip the heap back and forth between bins until the compost is ready, then you can store it one bin and use the other bin to start building the next heap.

This design facilitates turning with its removable divider between the bins. Simply slide the divider up and out of the way for easy shoveling. The front sides of the bins are full-width gates, providing easy access to the bins for moving material in or out.

But perhaps the best feature of this composter has nothing to do with production; it's all about appearances. As much as self-sufficient homeowners and gardeners love the idea of composting, few can honestly say they like the look of a compost heap. (And the aesthetics of plastic barrels or trashcan composters need no further criticism.) You may not see a lot of bin-type composters with cedar pickets, decorative posts, and traditional gates, but what would you rather look at: a pile of rotting garbage or a well-built picket fence?

TIP

Three-peat Enough

A three-bin system uses the same idea as a two-bin composter, but the additional bin helps make the process even more continuous. Once your heap is ready, flip it into bin two and use bins two and three for turning. This leaves bin one open for compiling the next heap. You can easily adapt this two-bin composter design to create a three-bin version. Just extend the overall length by a third, and create another center divider and gate. The two stringers along the backside of the structure can be cut from 12' 2 × 4s.

Composting is key in a truly self-sufficient garden, and the only thing better than an active composting bin, is a doubly active bin.

Building a Two-Bin Composter

CUTTING LIST

Key	No.	Part	Dimension	Material*
A	6	Posts	3½ × 3½ × 60"	4 × 4 PT pine
B	49	Pickets	¾ × 3½ × 36"	1 × 4 PT pine
C	2	Rear rail	1½ × 3½ × 93"	2 × 4 PT pine
D	4	Side rail	1½ × 3½ × 45"	2 × 4 PT pine
E	4	Hinge blocks	1½ × 3½ × 5"	2 × 4 PT pine
F	4	Gate rails	1½ × 3½ × 39¼"	2 × 4 PT pine
G	2	Latch block	1½ × 3½ × 3½"	2 × 4 PT pine
H	4	Divider panel stops	1½ × 1½ × 30"	2 × 2 PT pine
J	2	Divider rail	1½ × 3½ × 34"	2 × 4 PT pine

*Use pressure-treated lumber rated for ground contact or all-heart cedar, redwood, or other naturally rot-resistant species.

TOOLS & MATERIALS

4' level
Posthole digger
Miter saw
Cordless drill and bits
Gravel
Deck screws 1⅝", 3½"
Exterior-grade construction adhesive
½ × ½" galvanized hardware cloth with
    staples (optional)
(4) 3½ gate hinges with screws
(2) Gate latches with screws
4 × 4 post caps (optional)
Chalk
Framing square
Eye and ear protection
Work gloves

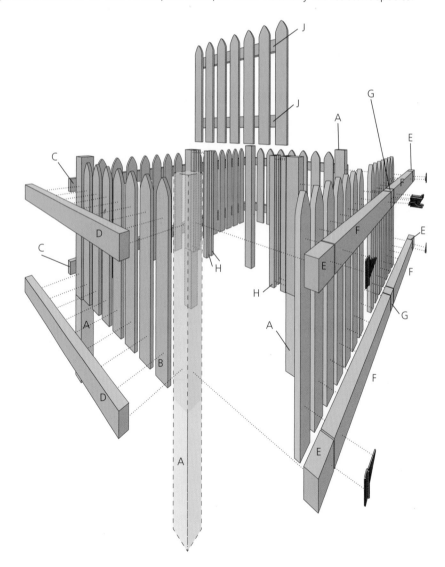

Choose a flat, level site at least 6 × 10' that allows for movement in front of the bins. Cut the back rails and posts to length, then lay them out on a flat surface. Square the assembly by making sure the diagonals are equal. Fasten the rails to the posts with 3½" deck screws.

Set the assembled back wall in place, then mark the hole locations with chalk or a shovel. Dig holes 8" in diameter by 24" deep at each post location.

Plumb and level the posts using wood braces. Fill the holes with gravel and dirt. The bottom rails should be roughly 6" to 8" above ground level.

(continued)

Building a Two-Bin Composter (continued)

Dig holes for the front posts, using a framing square and a side rail (or the 3-4-5 method) to locate the correct position. Put the posts in and fasten the top rail to both front and rear posts to help hold the front post plumb and level. Fill the holes with alternating layers of gravel and dirt and attach the bottom rails. Position the center post so that it's the same distance from the corner posts and in line with both of them.

At the front corner posts and front center post, measure and cut six short 2 × 4s to cover the front faces of the posts (5" long at the corners and 3½" at the center). These continue the runs of the stringers and will serve as mounting blocks for the gate hinges and latches. Install the blocks as you did with the rear stringers, but predrill and countersink all screw holes to avoid cracking the short pieces of wood.

Install the pickets along the back and sides with 1⅝" deck screws. Keep the pickets 1½" to 2" above the ground to prevent rot. The fastest way to install the pickets is to make spacer blocks. For this design, the seven pickets on the sides were spaced 1⅝" apart and the back and front pickets were spaced 1⅞" apart. To find the spacing for a different size bin, just subtract the total width of seven pickets (or however many you use) from the distance between the posts, then divide the result by 8 (the number of spaces between pickets). Use a 5"-high block of wood at the top to quickly establish the height for each picket.

Begin constructing the gates. Set the rails on your work surface so they are parallel and spaced the same distance apart as the bin stringers. Space the pickets 1⅞" apart like the back wall, but start from the center—line up the center of the first picket with the center of the rail, then work to each side so you end up with roughly 1⅞" between the last picket and the post. Fasten the pickets to the rails with 1⅝" screws and construction adhesive. The construction adhesive helps prevent the gate from sagging over time. Check the assembly for square as you work.

Clamp or screw a straight piece of wood across the three front posts to support the gates while the hinges and latches are attached. Hang the gates using gate hinges—both gates open out and away from the center post. Install latch hardware for each gate so it locks closed at the center post.

Cut the four stops for the sliding divider panel from 2 × 2s. Make marks at 1¼" in from the edges on both posts. Install the 2 × 2 stops against these lines with predrilled 3½" screws, creating a 1" slot at the center. The outside edges of the stops will overhang the edges of the posts by about ¼".

Build the divider panel using the same construction techniques used for the gates, but extend the outside pickets beyond the 2 × 4 rails so they fit into the channels between the stops. Space the pickets 2⅛" apart. The total width of the panel should be about ½" narrower than the distance between the posts so that the panel can slide in and out without binding. To keep the divider panel at the same height as the rest of the enclosure, screw in a small wood stop or a few screws near the bottom of the slot.

TIP

Upgrades

Add a touch of beauty to your compost bin by attaching a decorative post cap to the top of each 4 × 4 post. You can find a wide variety at home centers or online. You also may want to fasten a 2'- wide strip of ½" hardware cloth down to the ground around the inside of the composter to help keep all the compost inside the bin.

basement vermiculture bin

If you're new to vermiculture, this multi-bin unit is a great way to enter the wonderful world of worm poop. And if you've been working with a simple one-bin system in a plastic storage container, you're probably ready to take your composting to another level—or how about three levels? Actually, you can have as many levels as you like; they're all interchangeable and very easy to build. Just add bins as needed to increase production.

Multi-bin, or flow-through, vermiculture systems offer a few advantages over single-compartment composters. The main benefit is that flow-through systems allow you to harvest the compost without having to sift through the bin, or to dump out everything then put back the worms, bedding, and incomplete compost material. Harvesting is much easier for you and also less disruptive to the worms, helping them stay in "production" mode. The stacked design of vertical flow-through systems makes these composters space-efficient, an obvious plus when composting indoors.

The process couldn't be simpler: fill the upper bins with worms and compost material. As the worms eat the material, their castings (also known as vermicompost) fall through to the bottom bin. To harvest the compost, lift off the upper bins and empty the bottom level. Then reset the stack so the middle bin is now on the bottom, and the empty bottom bin is now on top; fill this bin with new material to keep the process going. The worms will travel up through the bins as they exhaust their food sources, so you never have to move more than a few stragglers at harvest time.

Vermiculture creates moisture (called leachate), which you'll have to catch in a tray or baking pan. You can set the stack of bins right into a pan, using blocks to ensure airflow, or you can build a handy base that supports the bins and provides a space for a tray to slide underneath (see Building the Optional Base, right). Speaking of moisture, if you're used to composting with a plastic tub, you'll find that an open wood system like this dries out more quickly than a plastic container, so be sure to check the moisture level regularly and mist the compost material with a spray bottle and non-chlorinated water as needed.

Building the Optional Base

The base is constructed much like the bin frames, with two ends fitting between two sides. Cut two 2 × 6 sides to 23½" long and two ends to 20½." Make a side cutout in the top edge of each 2 × 6, using a jigsaw. The cutouts allow you to grab the 1 × 2 edging to lift off the bins. Assemble the frame with 3" deck screws. Slide a sheet pan or plastic tray inside the base to collect leachate.

Make the most of worm castings (better known as poop), by setting up your own worm bin in the corner of a basement, or finish it with paint and keep it outdoors.

Building a Basement Vermiculture Bin

CUTTING LIST

Key	No.	Part	Dimension	Material
A	6	End	¾ × 3½ × 18½ "	1 × 4 clear pine or cedar
B	6	Side	¾ × 3½ × 20"	1 × 4 clear pine or cedar
C	6	End edging	¾ × 1½ × 20"	1 × 2 clear pine or cedar
D	6	Side edging	¾ × 1½ × 21½"	1 × 2 clear pine or cedar
E	1	Lid	¾ × 22 × 22"	¾" plywood
F	1	Handle	¾ × 1½ × 6"	1 × 2 clear pine or cedar

TOOLS & MATERIALS

Miter saw or circular saw
Cordless drill and bits
Wire cutters
Deck screws 1¼", 1", 2"
Waterproof wood glue
Galvanized hardware
  cloth with ¼" grid
  (20 × 60" min.)
¾" washer-head wood
  screws
Handle with screws
Drip tray and blocks
  (as applicable)
Sandpaper
Eye and ear protection
Work gloves

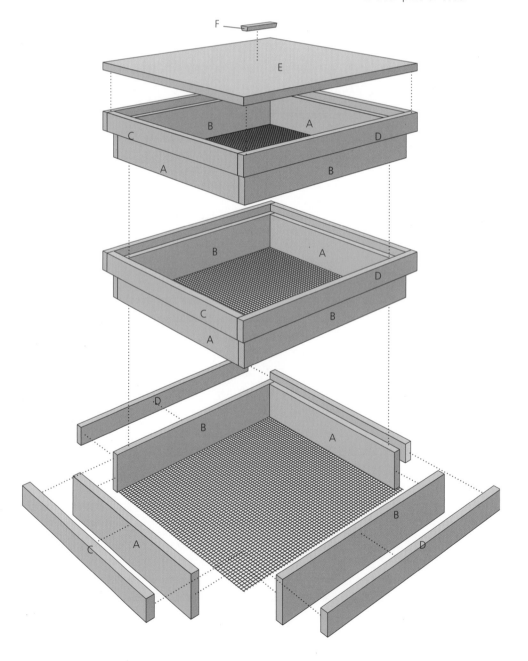

Assemble each bin frame by fitting the ends between the sides so all edges are flush. Drill pilot holes for three screws at each joint. Apply glue and fasten the joints with 2" deck screws. Use a square or measure diagonally between opposing corners to ensure the frame is square. Let the glue dry as directed.

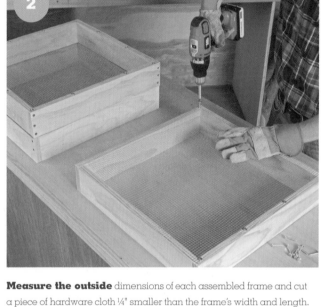

Measure the outside dimensions of each assembled frame and cut a piece of hardware cloth ¼" smaller than the frame's width and length. Center the mesh over the frame, leaving a ⅛" margin along all edges. Fasten the mesh to the frame edges with ¾" washer-head screws spaced about 4" apart.

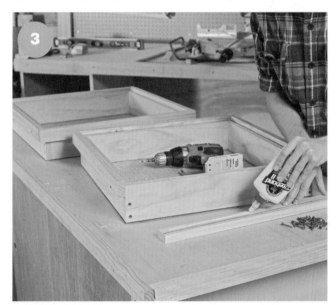

Add the 1 × 2 edging that serves as handles and ledges so that the bins can nest into one another. Position the edging against one side so the top is ½" above the frame's top edge, and the edging is flush with the frame ends. Drill pilot holes, apply glue, and fasten the edging to the frame with 1¼" deck screws. Install the rest of the edging in the same fashion.

Sand the edges of the lid as necessary and add a handle to the center of the panel. **Note:** Oil the wood parts of your bins to add moisture protection and to help keep the wood from absorbing liquids from the compost. Use a food-grade oil, such as walnut oil, and apply as directed.

soil sifter

10

Sifting your soil is an excellent way to refine the foundation of your garden. The basic idea is to sift the soil through a screen much as you would sift ingredients for baking. Sifting "cleans" the soil, removing large organic objects such as rocks and debris like broken glass. The process improves the texture of the soil, loosening it to allow for better water and air penetration. It can also remove old weed rhizomes—root systems that could grow new colonies of weeds. The benefits include improved drainage and moisture retention so that your plants' roots are more likely to get the water they need without becoming waterlogged or rotting.

You can take the opportunity of sifting your soil to blend in amendments such as compost, manure, or other nutritional additions. It's a great way to create a premium top soil that will get your garden off to a great start—and keep it growing strong throughout the season.

Sifting soil can be done with nothing more than a sturdy, thick mesh screen held by the edges. But if your garden is like most, you'll be faced with sifting quite a bit of soil and a simple hand-held screen will be quite laborious to use. That's why the design of the sifter described in the pages that follow is a bit more sophisticated. It uses a sifting box equipped with wheels, and this box sits in a frame. You sift the soil by rolling the box back and forth within the frame, saving a lot of energy, effort, and sore backs. If you want to make the rig even handier and easier to store, add handles to both the sifting box and frame.

The sifting frame has been sized to fit perfectly over a standard wheelbarrow. But if you are using another container to catch the sifted soil, or if your wheelbarrow is a different size, adjust the measurements to suit. This could even be used over an empty garbage can or barrel. Once you've constructed the sifter, sift soil for your whole garden, container plants, or anywhere you want clean, effective top soil. Your plants will thank you.

Sifting soil is largely a lost craft in the garden, but one that can go far toward improving your soil and making your plants grow as healthy as possible.

Building a Soil Sifter

Key	No.	Part	Dimension	Material
A	2	Frame Stile	¾ × 2½ × 35 "	1 × 3
B	2	Frame Rail	1½ × 3½ × 30 "	2 × 4
C	2	Frame Guide	¾ × ¾ × 35 "	¾ × ¾
D	2	Box Side	1½ × 3½ × 25 "	2 × 4
E	2	Box End	1½ × 3½ × 28 "	2 × 4

TOOLS & MATERIALS

(4) 1 " rigid casters (uni-directional)
¼" or ½" galvanized screen
Cordless drill and bits
1¼" washer head screws
  (also known as lath screws)
2½" deck screws
1¼" wood screws
1½"-wide metal angle
¾" × #8 pan head screws
Eye and ear protection
Work gloves

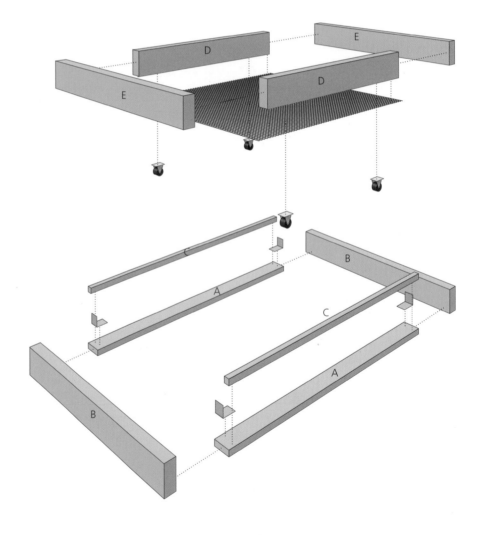

Drill pilot holes through the frame guides and into the 1 × 3 frame stiles. Screw the guides to the stile with 1¼" wood screws, ensuring that the guides are aligned along one edge of each stile. These guides will serve as tracks for the soil-sifting box.

Join the frame rails to the stiles with a metal angle at each corner.

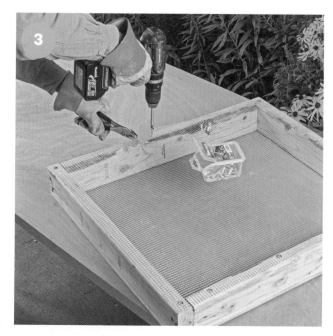

Screw the sifting box ends to the box sides with 2½" deck screws. Cut the screen ¼" less than the size of the box. Screw it to one side with washer head screws, then stretch it tightly and screw it to the opposite side. Use at least 4 screws per side.

Screw the casters to the back and front ends of each box side so that the wheels face toward the ends.

collecting rainwater

Practically everything around your house that requires water loves the natural goodness that's provided with soft rainwater. With a simple rain barrel, you can collect rainwater to irrigate your garden or lawn, water your houseplants, or top off swimming pools and hot tubs. A ready supply of rainwater is also a reliable stand-by for emergency use if your primary water supply is interrupted.

Collecting rainwater runoff in rain barrels can save thousands of gallons of tap water each year. A typical 40 × 40-foot roof is capable of collecting 1,000 gallons of water from only 1 inch of rain. A large rainwater collection system that squeezes every drop from your roof can provide most—or sometimes all—of the water used throughout the home, if it's combined with large cisterns, pumps, and purification processing.

Sprinkling your lawn and garden can consume as much as 40 percent of the total household water use during the growing season. A simple rain barrel system that limits collected water to outdoor (nonpotable) use only, like the rain barrels described on the following pages, can have a big impact on the self-sufficiency of your home, helping you save on utility expenses and reducing the energy used to process and purify water for your lawn and garden. Some communities now offer subsidies for rain barrel use, offering free or reduced-price barrels and downspout connection kits. Check with your local water authority for more information. Get smart with your water usage, and take advantage of the abundant supply from above.

NOTE: The collection of rainwater is restricted in some areas. Check with your municipality if you are unsure.

Rainwater that is collected in a rain barrel is free of the chemical additives and minerals usually found in tap water. This soft, warm (and free) water is perfect for gardens, lawns, and indoor plants like orchids that don't do well with tap water.

Rain Barrels

Rain barrels, either built from scratch or purchased as a kit, are a great way to irrigate a lawn or garden without running up your utilities bill. The most common systems include one or more rain barrels (40 to 80 gallons) positioned below gutter downspouts to collect water runoff from the roof. A hose or drip irrigation line can be connected to spigot valves at the bottom of the rain barrel. You can use a single barrel, or connect several rain barrels in series to collect and dispense even more rainwater.

Plastic rain barrel kits are available for purchase at many home centers for around $100. If kit prices aren't for you, a rain barrel is easy to make yourself for a fraction of the price. The most important component to your homemade barrel is the drum you choose.

Obtaining a Barrel

Practically any large waterproof container can be used to make a rain barrel. One easily obtained candidate is a trash can, preferably plastic, with a snap-on lid. A stan-dard 32-gallon can will work for a rain barrel, but if you can find a 44-gallon can, choose it instead. Although wood barrels are becoming more scarce, you can still get them from wineries. A used 55-gallon barrel can be obtained free or for a small charge from a bulk food supplier. Most 55-gallon barrels today are plastic, but some metal barrels are still floating around. Whatever the material, make sure the barrel did not contain any chemical or compound that could be harmful to plants, animals, or humans. If you don't know what was in it, don't use it. Choose a barrel made out of opaque material that lets as little light through as possible, reducing the risk of algae growth.

A barrelful of water is an appealing breeding ground for mosquitoes and a perfect incubator for algae. Filters and screens over the barrel opening should prevent insect infestation, but for added protection against mosquitoes add one tablespoon of vegetable oil to the water in the barrel. This coats the top surface of the stored water and deprives the larvae of oxygen.

TOOLS & MATERIALS

Barrel or trash can

Drill with spade bit

Jigsaw

Hole saw

Barb fitting with nut for overflow hose

1½" sump drain hose for overflow

¾" hose bibb or sillcock

¾" male pipe coupling

¾" bushing or bulkhead connector

Channel-type pliers

Fiberglass window screening

Cargo strap with ratchet

Teflon tape

Silicone caulk

How to Make a Rain Barrel

Cut a large opening in the barrel top or lid. Mark the size and shape of your opening—if using a bulk food barrel, mark a large semi-circle in the top of the barrel. If using a plastic garbage can with a lid, mark a 12"-dia. circle in the center of the lid. Drill a starter hole, and then cut out the shape with a jigsaw.

Install the overflow hose. Drill a hole near the top of the barrel for the overflow fitting. Thread the barb fitting into the hole and secure it to the barrel on the inside with the retainer nut and rubber washer (if provided). Slide the overflow hose into the barbed end of the barb elbow until the end of the hose seats against the elbow flange.

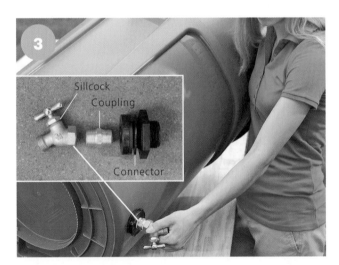

Drill the access hole for the spigot (either a hose bibb or sillcock, brass or PVC). Tighten the stem of the sillcock onto a threaded coupling inserted into the access hole. Inside the barrel, a rubber washer is slipped onto the coupling end and then a threaded bushing is tightened over the coupling to create a seal. Apply a strip of Teflon tape to all threaded parts before making each connection. Caulk around the spigot with clear silicone caulk.

Screen over the opening in the top of the barrel. Lay a piece of fiberglass insect mesh over the top of the trash can and secure it around the rim with a cargo strap or bungee cord that can be drawn drum-tight. Snap the trash can lid over the top. Once you have installed the rain barrel, periodically remove and clean the mesh.

How to Install a Rain Barrel

Whether you purchase a rain barrel or make your own from scratch or a kit, how well it meets your needs will depend on where you put it and how it is set up. Some rain barrels are temporary holding tanks that store water runoff just long enough to direct it into your yard through a hose and drip irrigation head. Other rain barrels are more of a reservoir that supplies water on-demand by filling up watering cans or buckets. If you plan to use the spigot as the primary means for dispensing water, you'll want to position the rain barrel well off the ground for easy access (raising your rain barrel has no effect on water pressure).

In addition to height, other issues surrounding the placement of your rain barrel (or rain barrels) include the need to provide a good base, orientation of the spigot and overflow, the position relative to your downspouts, and how to link more than one rain barrel together. **NOTE:** If you live in a cold climate, it's a good idea to drain the rain barrel and close it for the season to avoid having it crack when the water freezes.

TOOLS & MATERIALS

Drill/driver

Screwdriver

Hacksaw

Rainbarrel

Hose & fittings

Base material (pavers)

Downspout adapter and extension

Teflon tape

Sheet metal screws

Eye and ear protection

Work gloves

Select a location for the barrel under a downspout. Locate your barrel as close to the area you want to irrigate as possible. Make sure the barrel has a stable, level base.

Install the spigot. Some kits may include a second spigot for filling watering cans. Use Teflon tape at all threaded fittings to ensure a tight seal. Connect the overflow tube, and make sure it is pointed away from the foundation.

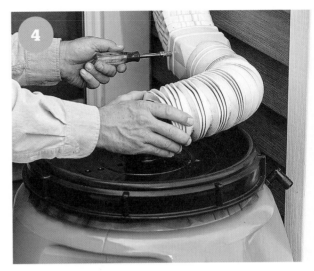

Cut the downspout to length with a hacksaw. Reconnect the elbow fitting to the downspout using sheet-metal screws. Attach the cover to the top of the rain barrel. Some systems include a cover with porous wire mesh, to which the downspout delivers water. Others include a cover with a sealed connection (next step).

Link the downspout elbow to the rain barrel with a length of flexible downspout extension attached to the elbow and the barrel cover.

Variation: If your barrel comes with a downspout adapter, cut away a segment of downspout and insert the adapter so it diverts water into the barrel.

Connect a drip irrigation tube or garden hose to the spigot. A Y-fitting, like the one shown here, will let you feed the drip irrigation system through a garden hose when the rain barrel is empty.

If you want, increase water storage by connecting two or more rain barrels together with a linking kit, available from many kit suppliers.

channeling rainwater

12

An arroyo is a dry streambed or watercourse in an arid climate that directs water runoff on the rare occasions when there is a downfall. In a home landscape an arroyo may be used for purely decorative purposes, with the placement of stones evoking water where the real thing is scarce.

Or it may serve a vital water-management function, directing storm runoff away from building foundations to areas where it may percolate into the ground and irrigate plants, creating a great spot for a rain garden. This water management function is becoming more important as municipalities struggle with an overload of storm sewer water, which can degrade water quality in rivers and lakes. Some communities now offer tax incentives to homeowners who keep water out of the street.

When designing your dry streambed, keep it natural and practical. Use local stone that's arranged as it would be found in a natural stream. Take a field trip to an area containing natural streams and make some observations. Note how quickly the water depth drops at the outside of bends where only larger stones can withstand the current. By the same token, note how gradually the water level drops at the inside of broad bends where water movement is slow. Place smaller river-rock gravel here, as it would accumulate in a natural stream.

Large heavy stones with flat tops may serve as step stones, allowing paths to cross or even follow dry stream beds.

The most important design standard with dry streambeds is to avoid regularity. Stones are never spaced evenly in nature and nor should they be in your arroyo. If you dig a bed with consistent width, it will look like a canal or a drainage ditch, not a stream. And consider other yard elements and furnishings. For example, an arroyo presents a nice opportunity to add a landscape bridge or two to your yard.

An arroyo is a drainage swale lined with rocks that directs runoff water from a point of origin, such as a gutter downspout, to a destination, such as a sewer drain or a rain garden.

Building an Arroyo

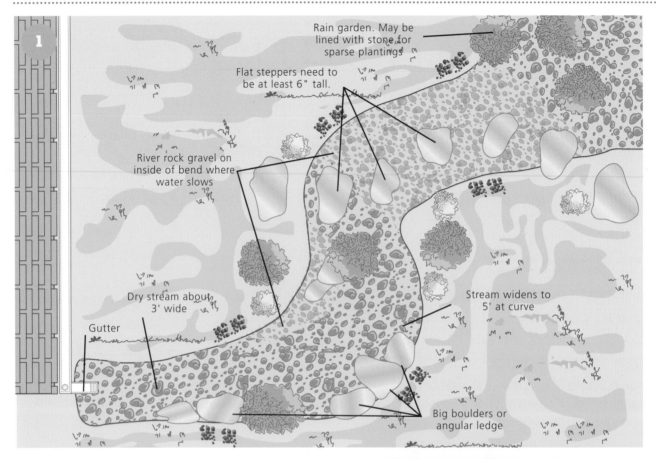

1

Rain garden. May be lined with stone, for sparse plantings

Flat steppers need to be at least 6" tall.

River rock gravel on inside of bend where water slows

Dry stream about 3' wide

Gutter

Stream widens to 5' at curve

Big boulders or angular ledge

Create a plan for the arroyo. The best designs have a very natural shape and a rock distribution strategy that mimics the look of a stream. Arrange a series of flat steppers at some point to create a bridge.

TOOLS & MATERIALS

Landscape paint
Carpenter's level
Spades
Garden rake
Wheelbarrow
Landscape fabric or 6-mil black plastic
Mulch
8"-thick steppers
6 to 18" dia. river-rock boulders
¾ to 2" river rock
Native grasses or other perennials for banks
Eye and ear protection
Work gloves

2

Lay out the dry stream bed, following the native topography of your yard as much as possible. Mark the borders and then step back and review it from several perspectives.

Excavate the soil to a depth of at least 12" (30cm) in the arroyo area. Use the soil you dig up to embellish or repair your yard.

Rake and smooth out the soil in the project area. Install an underlayment of landscape fabric over the entire dry stream bed. Keep the fabric loose so you have room to manipulate it later if the need arises.

Place flagstone steppers or boulders with relatively flat surfaces in a stepping-stone pattern to make a pathway across the arroyo (left photo). Alternately, create a "bridge" in an area where you're likely to be walking (right photo).

(continued)

Building an Arroyo (continued)

Add more stones, including steppers and medium-size landscape boulders. Use smaller aggregate to create the stream bed, filling in and around, but not covering, the larger rocks.

Dress up your new arroyo by planting native grasses and perennials around its banks.

What Is a Rain Garden?

A rain garden is simply a shallow, wide depression at least 10' away from a basement foundation that collects storm water runoff. Rain gardens are planted with native flood-tolerant plants and typically hold water for only hours after rainfall. Check your local garden center or extension service to find details about creating rain gardens in your area.

Alternative: Create a Swale

A swale is basically just a gently sloping trench filled with grass or plants instead of stone. Like an arroyo, a swale channels runoff from a gutter or low area to a rain garden or other section of your yard. Use stakes or spray paint to mark a swale route that directs water away from the problem area toward a runoff zone.

Remove sod and soil from the marked zone. Cut the sod carefully (or rent a sod cutter) and set it aside to reuse if it's in good condition.

Level the trench by laying a 2 × 4 board with a carpenter's level on the foundation. Distribute soil so the base is even, and so that it slopes enough to move the water in the direction you want it to go.

Lay sod in the trench to complete the swale. Compress the sod and water the area thoroughly to check drainage.

sub-irrigated planter

13

Sub-irrigated planters are a great way to keep plants strong and healthy without spending a lot of time watering and without worrying whether you're soaking the roots deeply enough. With a sub-irrigated planter (a.k.a. "SIP") plants are watered from the bottom up, drawing moisture as they need it from a large reservoir of water deep below the surface. Instead of just saturating the top few inches of dirt every few days and then having most of it evaporate away, water is poured into the reservoir through an intake pipe and then wicked upward to the roots, resulting in less evaporation and a more consistent water supply—one that can last for weeks.

Here's how it works: First, waterproof the planter or pot with thick, 6-mil PVC, rubber pond liner, or, if you're using a large pot, by plugging the holes with epoxy or caulk. You can turn anything from a large raised bed to a regular plastic pot into a sub-irrigated planter, as long as you seal the bottom so the water doesn't drain away. For this project we used 4 × 6 timbers because they're simple, last forever, and matched an existing planter elsewhere on the patio, but you could easily substitute 2 × 6s or 2 × 8s. Just reinforce the sides with a few vertical 2 × 2s to keep the boards from bowing out.

Form the reservoir with perforated drain tile wrapped with a drain tile sock (which keeps the tile from filling with dirt) with a fill tube at one end and an outlet at the other. The soil packed around the drain tile wicks out the moisture, and capillary action pulls it up into the higher layers of soil. The air space in the drain tile also aerates the soil under the plants, which makes the soil and the plants healthier. You don't have to worry about overwatering or turning the soil in the planter into a big mud puddle because the excess water drains through the outlet when it reaches the top of the drain tile, even if the source of the water is a downpour.

Keep your plants happy and fresh, even in scorching weather, with a sub-irrigated planter, which provides abundant moisture to roots rather than just the top few inches of soil.

Building a Sub-Irrigated Planter

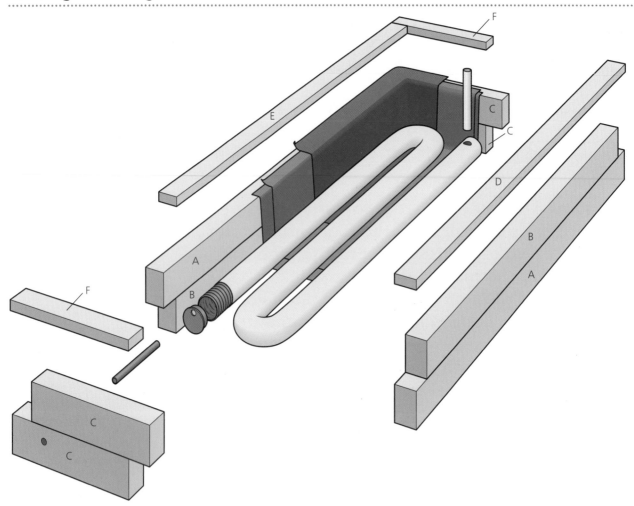

TOOLS & MATERIALS

Drill
Impact driver (optional)
Razor knife
Scissors
Stapler
Miter saw or circular saw
⅝" spade bit with extension
1½" hole saw

(22) 8" exterior, self-tapping ⅜" screws
(3) 8' × 4" perforated flexible drain tile
(2) 4" drain tile caps
4" drain tile coupler
1½" × 12" PVC tube
½" × 12" drain tube
Weed barrier
6-mil poly

Staples
3" deck screws
Potting mix
(5) 4 × 6 × 8" pressure treated
(3) 2 × 4 × 8" pressure treated
Eye and ear protection
Work gloves

CUTTING LIST

Key	No.	Part	Dimension	Material
A	2	Side	3½ × 5½ × 90"	4 × 6 P.T.
B	2	Side	3½ × 5½ × 83"	4 × 6 P.T.
C	4	End	3½ × 5½ × 20½"	4 × 6 P.T.
D	1	Front cap	1½ × 3½ × 90½"	2 × 4 P.T.
E	1	Rear cap	1½ × 3½ × 83½"	2 × 4 P.T.
F	2	Side cap	1½ × 3½ × 20 ¾"	2 × 4 P.T.

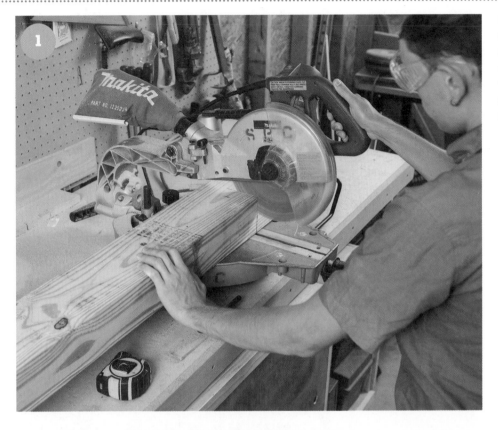

Cut the 4 × 6s to length on a miter saw, or make two passes with a circular saw. Most miter saws can cut a 4 × 6, but be sure to support the ends of the heavy 4 × 6. Check to make sure the factory ends are cut square before measuring.

Assemble the planter with 8" self-tapping bolts. Overlap the corners on the second layer for extra strength and rigidity so that the frame resists the outward pressure of the heavy soil. Also drive bolts down through the second layer of 4 × 6 into the first layer at the corners and at the centers of the long sides. Build the planter near the final location—it will be very difficult to move once it's together.

(continued)

Building a Sub-Irrigated Planter (continued)

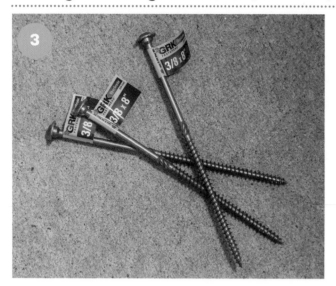

Assemble the planter with self-tapping, exterior-grade construction screws instead of nails or traditional lag screws. They're made from harder steel, and even large screws such as these can be driven in without predrilling thanks to the self-tapping feature, though it helps to use an impact driver.

Remove any protruding stones or roots from the bottom of the planter, then spread two layers of weed barrier over the bottom and up the sides of the planter. This will help prevent punctures in the 6-mil poly lining of the planter. Spread it loosely and trim back the excess and folds around the top. If you build this planter on a deck, build it on a wood base with an air gap between the deck wood and the planter wood to prevent damage if a leak ever develops.

Spread the poly out, gently tucking it into the corners. Leave it loose so that it can expand as it's filled with soil and water. Fold it over the 4 × 6 and staple it at the top edge only. Trim the edges so they'll be hidden by the 2 × 4 cap. If your planter will be large or you'll be working the soil a lot, consider using a rubber fish pond liner, which is more durable and puncture-resistant.

Add the 2 × 4 top cap, fastening the pieces with 3" deck screws. Overlap the corners, log-cabin style, to lock all the pieces together. We cut the 2 × 4s slightly longer than the 4 × 6s to create a slight overhang and shadow line.

Connect the lengths of tubing, covering them with a filter sock if needed. The type we bought could be lengthened or shortened accordion-style and the pieces just snapped together. Lay it in the planter and plug each end with a cap (you may need a transition at one end) or just jam it tightly against the PVC.

Cut a hole near the beginning of the hose for the inlet pipe, which should be 1" to 1½" O.D. PVC or ABS—big enough for a hose to fit in. The inlet pipe should stick up about 2" to 4" above the soil level. Use a hole saw (carefully!) or a razor knife to cut the hole.

The overflow tube should be about half the size of the inlet and located near the top of the drain tile at the opposite end of the hose. Measure the spot, then drill through the 4 × 6 using a 12" drill bit the diameter of the tube. Drill carefully and slowly. Stop drilling when the tip of the bit pokes through the wood so you can make a small slit in the poly and push the bit through. If you try to drill straight through the poly, it may become twisted and torn and need to be patched. Locate the correct spot at the top of the plastic cap and slowly drill through it. Cut the tubing about 10" long and work it through the hole into the drain tile cap. You may need to ream the hole in the wood a little to make room for the tube.

Fill the drain tile to make sure the overflow tube works right and to check for leaks—better to know now, when you can still do something about it. If you discover a leak, remove the drain tile and add another layer of poly. If there are no leaks, add the soil. A potting mix works best. After you've filled the planter with soil, add more water until it starts to come out the drain hose. Then add your plants. Check the water level at the bottom of the fill tube every few days and add more when it gets low.

garden projects

Most of us don't have acreage to plant; instead, we're working with relatively small urban and suburban lots where planting space is limited, where vegetables and flower gardens have to share space with kids, yard furniture, decks, barbecues, shade trees, and pets. But limited space doesn't mean you have to limit your ambitions. The trick is to be creative with the space you have, and to grow more with less.

In this section we'll show you how to expand your growing space and the size of your crop by growing more efficiently and by growing upward—where the space is virtually unlimited. Training plants to grow on trellises like the ones shown in this section will substantially increase yields of tomatoes, beans, peppers, and other vegetables by giving them more room and light and by keeping them off the ground and away from mold and insects. Cold frames increase your yield by extending the growing season, allowing you to start planting in late winter or early spring. Raised beds let you transform ordinary, tired dirt into state-of-the-art soil. Instead of trying to work infertile, rocky dirt, you just build a box on top of it and fill it with a rich, high-quality growing medium designed to produce high yields.

starting & transplanting seedlings

14

Add weeks to your garden's growing season by starting seeds indoors, then transplanting them to the garden after the danger of frost is past. Seedlings are available for purchase at the garden center in the spring, of course, but starting your own at home presents a number of advantages:

- Buying seeds is less expensive than buying seedlings.

- You can cull out all but the strongest seedlings, which will hopefully result in stronger plants and a more bountiful crop.

- Garden centers sell seeds for a diverse and varied array of plants but seedlings for only the most common species. Seed catalogs introduce an entirely new selection as well.

- You can be certain that unwanted pesticides have not been used on the plants in your garden, even in their infancy.

Start your seeds 8 to 10 weeks before you plan to transplant them into your garden. To get started, you'll need a few small containers, a suitable growing medium, and a bright spot for the seedlings to grow—either a sunny window that receives at least six hours of bright sunlight per day, a greenhouse, or a planting table in your home that's

Using colored cups as starter containers has the advantage of letting you color-code your plants so you can avoid any confusion. Beginning gardeners often have trouble distinguishing one pot from another when they are still seedlings. Here, the red cups clearly communicate "tomato."

Seedlings need a lot of water, sunlight, and warmth in their infancy. A kitchen window or greenhouse is an ideal growing environment for propagating plants.

illuminated by artificial grow lights. If you're planning to raise your seedlings by artificial light, position one or two fluorescent lighting fixtures fitted with 40-watt, full-spectrum bulbs about 6 inches above the seedlings. Leave fluorescent lights on for 12 to 16 hours a day—many gardeners find it helpful to connect the fixture to a timer to ensure their plants receive adequate light each day.

A 4', two-bulb fluorescent light fixture that can be raised or lowered over a table is really all you need to start your vegetable plants indoors.

Growing Mediums

If you plan to use your own garden soil or compost, prepare your seedling containers in the fall, before the ground gets too cold and wet.

Almost any small container can be used to grow seedlings. Just make sure the container you use is clean and hasn't had contact with any chemical that could be poisonous to plants. Also, remember to cut a drainage hole in the bottom of your container before filling it with soil. Drainage is very important to ensure that your plants are well ventilated. Excessively moist soil can result in mold or other diseases, as well. Good options for seedling containers include:

- Peat pots or pellets
- Fiber cubes
- Used plastic jugs
- Cans (any size)
- Used plastic tubs (i.e., sour cream, cottage cheese, or margarine containers)
- Used yogurt cups
- Egg cartons
- Small paper cups

○ Starting Seedlings

1. **Planting:** Sow three to four seeds in each container according to the instructions on the seed packet—as a general rule, large seeds should be buried and small seeds can be sprinkled on top of the soil. Label the container with the type of plant and the date your seeds were planted.

2. **Germination:** Water the seeds whenever the containers look dry. Until the seeds sprout, keep seedlings in a dark, warm space. Cover germinating seeds with plastic bags or plastic wrap. Open the plastic for a few hours every few days to let the soil breathe, then re-close.

3. **Sprouts:** When the seeds sprout, remove plastic covering and move them into direct light. Seedlings need lots of light to grow. Keep the soil medium moist but not soggy. Remember, multiple light waterings are better for seedlings than the occasional soaking.

4. **Culling:** When the true leaves appear (see illustration), cut off all but the strongest seedling in each container at soil level. Do not pull up the unwanted seedlings, as this may damage the roots of the seedling you're cultivating. You may also choose to fertilize every week or so as your seedlings grow.

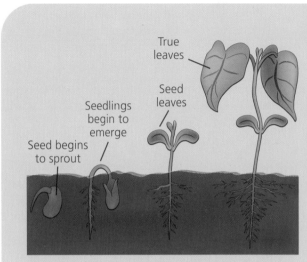

The four stages in the growth of a seedling are illustrated above. Note that seed leaves and true leaves serve different purposes and will look different. When a plant has four to eight true leaves, it is ready for transplanting.

○ Hardening Off & Transplanting

When your seedlings have four to eight true leaves, they should be hardened off, then transplanted to the garden. Hardening off is the process of gradually introducing the plant to outdoor conditions so it is not shocked when you move it outside permanently. About two weeks before planting, place your seedlings outdoors for an hour the first day, and then gradually increase the time until your seedlings spend the whole day outside. Protect seedlings from wind and do not expose them to the midday sun for the first few days. Stop fertilizing seedlings during the last week. A cold frame is a great environment for hardening off seedlings. Seedlings can stay in a cold frame for two to three weeks, gradually getting used to the cooler air and chilly nights before they go out into the garden. Open the lid of the cold frame a little bit more each day.

Transplant your seedlings into the garden on a cloudy day or in late afternoon to avoid excessive drying from the sun. Remove seedlings gently from their containers, holding as much soil as possible around the roots (containers that are pressed from peat are not intended to be removed). Place each into a hole in your garden, spreading the roots carefully, then pack soil around the seedling to hold it straight and strong. Thoroughly soak all seedlings with a very gentle water spray after they've been planted. If you have a rain barrel or another source for untreated water, this is a perfect application for it: the chlorine in most municipal water can be harmful to delicate plants.

potato-growing box

Potatoes have a funny way of growing. They start under soil, like most plants, but as their shoots reach up through the soil toward the sun, they develop roots that extend horizontally from the main taproot, ultimately yielding fruit. If you leave the emerged foliage alone, each plant will grow one batch of potatoes, but if you cover some of the foliage with soil, more roots will grow underground, yielding more fruit for harvest.

This space-efficient method of "vertical gardening" is made possible with a potato box. You plant your seed potatoes inside a single level, or course, of the box, covering them with soft soil or mulch. When the plants grow to a height of 8 to 12 inches above the soil, you add another layer to the box and cover about one-third of the plants' height with soil; the buried portions of foliage will form new lateral roots as the plant continues its upward climb. Repeat the process until harvest time, when you simply disassemble the box to get to your spuds.

This potato-box design includes two special features that facilitate assembly and disassembly, as well as off-season storage. The box sides are held together with half-lap joints, so no fasteners are needed for assembly. The boards on opposing sides of the course have 2 × 2 cleats that extend down to interlock with the course below to prevent shifting. At harvest, the box quickly disassembles without tools or digging, and the boards stack neatly for storage.

TIP

Growing Superlative Spuds

• Start with disease-free "seed potatoes" purchased from a garden center or seed store.
• Proper watering is key to a healthy, nice-looking crop. Maintain even moisture levels and monitor the box carefully because container-grown potatoes can dry out relatively quickly.
• Prevent disease contamination by moving your potato box to a new location each year and use new soil for each growing season.
• In climates with frost, harvest potatoes after frost kills the exposed foliage. In climates without frost, cut off the plants and let the potatoes sit for 1½ to 2 weeks before harvesting.
• Check with a local extension office or master gardener for potato growing and storage tips specific to your area, as well as recommendations for potato varieties that thrive in the local climate.

Growing a wealth of potatoes can be easy as long as you have the right container. Look no further for the new garden home of your next starchy crop.

Building a Potato-Growing Box

CUTTING LIST

Key	No.	Part	Dimension	Material
A	16	Box side	1½ × 5½ × 29½"	2 × 6
B	16	Cleat	1½ × 1½ × 5"	2 × 2

*All-heartwood redwood or cedar or other naturally rot-resistant lumber is recommended. Avoid cupped boards.

TOOLS & MATERIALS

Table saw or circular saw
Drill
2½" deck screws
Mallet
Chisel
Pencil
Square
Eye and ear protection
Work gloves

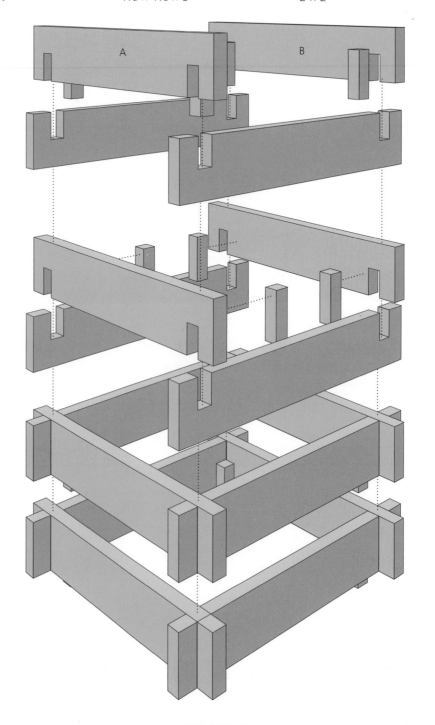

Cut the sides for the box carefully. All the pieces need to be exactly the same dimensions and the ends perfectly square. Lay out the half-lap notches on two of the boards as shown, making each 1⁹⁄₁₆" wide. Screw a straight board to the sliding miter gauge to keep the sides straight as you push them through. The depth of each notch is exactly half of the board's width, so measure the boards and adjust the depth as needed. Align and clamp the boards to the miter gauge. Make the first cut at 2" from the end, then slide the boards over 1⁹⁄₁₆" and make the second cut. (You can also use a circular saw or a jigsaw to make these cuts.)

Use a square to mark positioning lines for the cleats on each board, marking the lines just to the inside of each notch. These represent the outside faces of the 2 × 2 cleats. Mark a vertical line 3" above the base of each notch; this represents the top end of each cleat. Install the cleats by drilling countersunk pilot holes and fastening the cleats to the boards with pairs of 2½" deck screws. **NOTE:** For the bottom course of the box, you can omit the cleats, if desired, or you can make longer cleats with pointed ends to help anchor the box to the ground.

Break off the waste material and clean up the bottom of the notch with a sharp chisel. Check the fit. If everything looks good, cut the rest of the boards on the table saw. If you're using a circular saw, trace the notches onto the remaining boards with a sharp pencil, then cut out all of the remaining notches. **NOTE:** Only the bottom layer of boards has to be the exact same depth. If it's a little deep, you can drive a screw in at the base and turn it in or out to make the top edges level.

Test-fit the parts by assembling the entire box. Use a framing square or measure diagonally between opposing corners of the box frame to make sure the frame is square. Plant your seed potatoes about 4 to 8" deep and cover with soil or mulch. As the plants spring out of the ground, add more soil and additional courses until the box is complete.

octagon
strawberry
planter &
cover

16

Growing strawberries is equally popular among gardeners and animals. Gardeners love them because they grow like weeds in almost any climate, and there's no fruit better than a freshly picked organic strawberry. Animals like them for that second reason. The bright, red color and sweet fragrance of ripening strawberries is a siren song to every squirrel and other varmint within a mile radius, it seems. If you've grown strawberries in an open garden patch, you also know that the plants tend to multiply like rabbits, quickly taking over the patch if not controlled.

This easy-to-build planter covers you on both fronts. The wood frame attractively contains the plants while keeping them all within easy reach, and the lightweight mesh cover critter-proofs your crop in seconds. The planter looks best without the cover, so you can leave it off while the fruit is still green. Just be sure to keep the cover in place once the strawberries start to ripen (you can be sure the birds and other critters are watching … and waiting).

Building the planter frame is a cinch with a power miter saw, but you can also use a circular saw or even a handsaw and a tall miter box. Most miter saws will cut a 2 × 4 set on-edge; for larger lumber, you'll need a 12-inch miter saw or a sliding compound miter saw, which can make the miter cuts while the wood is flat on the saw table. You can also cut multiple 2 × 4s and stack them to create a taller planter.

The planter as shown is 69 inches wide. If desired, you can modify the size simply by cutting shorter or longer pieces; they're all the same size and all have their ends cut at 22½ degrees. You can even change the shape of the planter—to create a hexagon or decagon, for example. The math is simple: to determine the angle cuts, simply double the number of sides and divide that number into 360. For example, a hexagon has six sides; therefore: $360 \div 12 = 30$. Make each end cut at 30 degrees and you'll have a perfect hexagon. The mesh cover is custom-fit to the planter size. You'll find the bird netting we used for a cover at nurseries and home centers.

Option for Off-Season Storage

Leave off the glue and use metal strapping and screws on two of the wood joints so that you can disassemble the planter for winter. Choose any two opposing corner joints (dividing the octagonal frame in half) and join the wood members with metal straps, or ties, and 1½" galvanized screws. The metal straps are designed for structural framing connections and are sold at home centers. Check the building materials department for deck-framing connectors.

Strawberries are some of the most popular fruits grown in the home garden. This octagonal planter provides for as much as you can grow, eat, make into jam or put up.

Building an Octagon Strawberry Planter & Cover

CUTTING LIST

Key	No.	Part	Dimension	Material*
A	8	Planter side	1½ × 3½ × 28½"	2 × 4
B	4	Cover support	¼ × ¾ × 72"	Screen molding

*All-heart cedar or redwood or other naturally rot-resistant wood species is recommended.

TOOLS & MATERIALS

Miter saw
Cordless drill and bits
Waterproof wood glue

Mulch
3" deck screws
Bird block netting

Speed square
Eye and ear protection
Work gloves

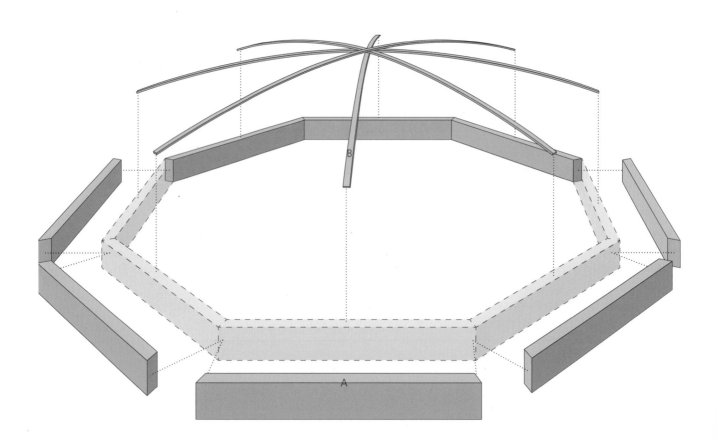

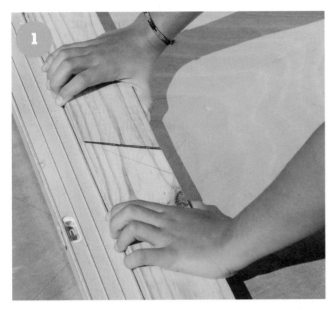

Set up your miter saw and run some test cuts to make sure the angle setting is accurate. Cut two pieces of scrap 2 × 4 at 22½°. Fit the pieces together end to end and check the assembly with a straight edge; the pieces should be perfectly straight. (If you're using a circular saw, cut test pieces with the saw blade tilted at 22½°.)

Use a stop block setup to ensure that all pieces are precisely the same length. This speeds your work and ensures the joints will fit tightly. For this roughly 6'-wide planter, each piece is 28½" long, measuring between the long points of the two angled ends. To make the cuts, leave the miter saw set on one side and just flip the board over for each successive cut.

Drill pilot holes through one frame piece and into its mating piece, starting the hole about ⅞" from the pointed end of the first piece. Angle the hole slightly toward the inside of the joint (what will be the inside of the assembled frame). This gives the screws a little bit of cross-grain penetration, helping them hold better than if they were perfectly parallel to the grain of the mating piece.

Apply glue and fasten each joint with two screws. For extra strength, add a third screw driven from the opposite direction as the first two. **TIP:** Work on a flat surface, such as a garage floor, to help keep the pieces flush at the top and bottom so the entire assembly will be flat.

(continued)

Building an Octagon Strawberry Planter & Cover (continued)

Dig out the grass and soil at least several inches deep and replace it with high quality garden soil mix. Cover the soil with landscape fabric to keep unwanted grass or weeds from growing. Use nails or spikes to hold the fabric in place, then trim it flush with the octagon edges.

Create an arched dome with wooden screen molding. Predrill holes at each end for screws, using a drill bit bigger than the size of the screw. Mark the centers of each 2 × 4. Screw one end in, leaving the screw about ½" out of the wood. Move to the opposite side, arch the wood, and put a screw in on that side, tightening it just enough to hold the end down. Join the frame rails to the stiles with a metal corner at each stile.

Spread an inch or two of mulch over the landcape fabric, then plant the octagon with strawberry plants. Cut slits in the fabric where each plant goes and then work the plant down into the garden soil.

Spread bird netting over the dome. Hold it in place by hooking it on two roundhead screws or small nails on each side. Trim the excess. To take the netting off, wrap one side around a piece of smooth wood or a 6'-long pole and just roll it up. When you want to cover the strawberries in the spring, just hook one side and roll it out again.

Project Detail: Filling Your Strawberry Planter

Strawberries are some of the most popular plants among home gardeners, and for good reason. Put in a little bit of work and diligence, and you'll be rewarded with a bumper crop of incredibly sweet and delicious fruit. Building the octagonal planter in this project is the first step to success—by growing in a raised planter, you control the growing culture and ensure protection for the plants. The next step is the soil.

Fill the planter with loamy soil rich in organic matter, with a pH around 6. Use bagged soil to avoid any problems; strawberries don't grow well in soil that has supported sod, and are susceptible to disease from soils in which peppers, tomatoes, eggplant, and potatoes have grown.

Next, you'll need to decide on the type of strawberries you want to plant (the type available in your local growing zone may influence your decision, so it's best to check catalogs and local nurseries before settling on a specific variety). You'll choose between three different types: June bearing, everbearing, and day neutral. June bearing strawberries are the largest and produce all their fruit over a three-week period (this type is further broken down into early, late, or midseason varieties).

Despite their name, everbearing strawberries don't produce continuous harvests. Instead, most will yield three separate harvests, roughly corresponding to spring, summer, and fall. Day neutral, on the other hand, will grow fruit all through the warm months. Most gardeners—looking to optimize their yields—grow more than one variety. However, everbearing and day neutral are better suited to the confines of a planter or raised bed.

Once you get your plants home, you'll need to plant them correctly. Many novice home gardeners plant their strawberry plants improperly, causing lower yields and even die-off between seasons (if properly cared for, your plants will produce strawberries for four years or more). Plant strawberries after all threat of frost has passed. Always plant on an overcast day, or early in the morning before the sun is bright. The shock of transplanting coupled with bright direct sun can damage strawberry plants.

Each plant should be buried deep enough so that the roots are just barely but completely covered, and the crown is exposed (see below).

This is the proper planting depth for a strawberry plant; notice that no portion of the roots is exposed and that the crown is fully aboveground.

Professionals recommeand removing the flowers on June-bearing strawberries when they appear in the first year. When the flowers are removed, the plant puts all its energy into growing strong roots and runners that will result in new plants. This translates to a much larger crop the second year. Nursery pros recommend removing the flowers on everbearing and day neutral varieties until the end of June for the same reason.

Aftercare is important for strawberry plant health. Make sure they are getting six full hours of sun a day. If not, move the planter. Before the next season, cut off old foliage down to the crown each year, and mulch the plants well over winter, to ensure against freezing.

vertical planter

17

Running out of gardening space? You can still grow a bumper crop of herbs and vegetables with a vertical planter, even if you only have a postage-stamp–sized balcony or patio. This stair-step planter gives you almost 7 square feet of planting area, but takes up less than 2 square feet of floor space. It also provides a useful storage nook underneath, where you can store supplies and garden equipment. Place it next to an entry door and you can just reach out and grab fresh herbs for dinner.

This project is made from standard lumber and inexpensive planters, and you can easily modify it to fit the space you have or old containers or pots you have lying around. To change the dimensions, all you need to do is change the rise and run and the distance between the stair stringers. You can also make the planter wider by adding additional stringers. You'll need a stringer about every 2 feet—so if you use a 36-inch planter instead of the 24-inch one we used, add a stringer in the center.

If you like the design but would rather use pots instead of rectangular planters, simply screw on 1× or 2× treads to the stringers and place the pots on those.

The key tool you'll need for building the planter is a full-size framing square. If you've never built stairs before, it may seem a little mysterious, but that's the only tool you'll need to make the planter steps perfectly level and plumb (assuming the planter is set on a reasonably level surface). The first step is to select a planter and measure it, or decide on the width and height of the step if you're using pots. Locate the width, which is called the "run," and the height (the "rise") on the square and mark those distances on the wide board (the "stringer") that the steps are cut from. Squaring off the top and bottom steps as we show in the photos makes the steps level. Then it's just a matter of supporting them with a post or screwing them to a wall.

Not a fan of carpentry? You can simplify this project by using pre-cut deck stair stringers instead of cutting your own.

Plant more in less space with a vertical planter. It's a great way to keep a fresh supply of herbs close to your kitchen.

Building a Vertical Planter

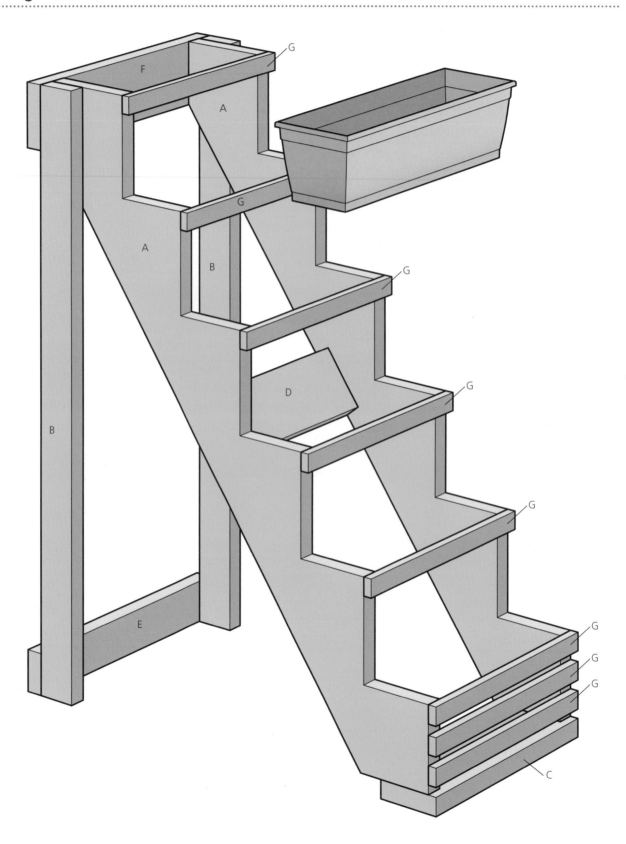

TOOLS & MATERIALS

Carpenter's framing square
Drill
Tape measure
Hammer
Circular saw
Jigsaw (or handsaw)
Miter saw (optional)

Deck screws or exterior-grade
  self-tapping screws—1¼", 2", 3"
Countersink drill bit
(4) ⁵⁄₁₆" × 3½" galvanized carriage bolts
  with washers and nuts
(6) 24"-long planter boxes
Large bag of potting mix or soil

(2) 2 × 10 × 8' pressure treated
(1) 2 × 4 × 12' pressure treated
(1) 2 × 6 × 8' pressure treated
(2) 1 × 2 × 8' pressure treated
Eye and ear protection
Work gloves

CUTTING LIST

Key	No.	Part	Dimension	Material
A	2	Stringer	1½ × 9¼ × 96" (cut to length)	2 × 10 P.T.
B	2	Post	1½ × 3½ × 51"	2 × 4 P.T.
C	1	Bottom cross support	1½ × 5½ × 20"	2 × 6 P.T.
D	1	Center cross support	1½ × 5½ × 17"	2 × 6 P.T.
E	1	Post footer	1½ × 3½ × 23"	2 × 4 P.T.
F	1	Post header	1½ × 5½ × 23"	2 × 6 P.T.
G	8	Riser	¾ × 1½ × 20"	1 × 2 P.T.

Building a Vertical Planter

Measure the top and bottom of the planter box style you purchase to determine the cuts for the stair step display. Use a framing square to determine the width if the planter box you chose has sloping sides.

Locate the measurements for the rise (8½") and run (6½") on the inside of the framing square (8½" on the narrow blade and 6 ½" on the wide blade). Hold the square so that the measurements intersect the edge of the 2 × 10, then mark along the inside edge of the framing square.

(continued)

Building a Vertical Planter (continued)

Cut the rise and run for each step with a circular saw, then finish the cut with a jigsaw or handsaw. Cut the top run at 6½" and the bottom rise at 7", squaring off the two ends as shown in the photo. Making the bottom rise shorter allows for a crosspiece, which protects the vulnerable end grain of the riser from rot.

Cut the top and bottom crosspieces for the rear support posts and screw them to the 51"-long 2 × 4 posts with 3" screws. Predrill or use self-tapping screws to avoid splitting the wood. Make sure the assembly is square before driving the final screws in. If your planters are a different height than ours, just add all the rises together to get the height of the post.

¾" mark

Space the stringers the proper distance by screwing on 2 × 6 cross supports at the bottom and center (20" and 17", respectively). Make sure the bottom 2 × 6 projects ¾" past the front of the stringers.

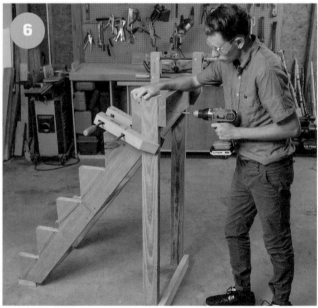

Stand the leg assembly upright, then lift and fit the step assembly into it. Fasten the stringers to the 2 × 6 cross support with 3" screws.

Make sure the steps are at least roughly level. Drill holes for the bolts, then tap the bolts through and tighten the nuts.

Cut all the 1 × 2 risers to length, then nail or screw them in place. You can use 6d galvanized casing nails, but 2" or 2½" self-tapping screws like the ones we use hold better and look attractive when they're all screwed in the same spot. Make a mark at the center of each 1 × 2, 1 inch from the end, then predrill with a bit slightly larger than the screw diameter.

Set the planters on the steps and align them with each other. Screw in the top edges with one self-tapping screw into each stringer. Leave the screws slightly loose. Place the completed project in a sunny spot outside and fill with soil and plants.

clothesline trellis

18

Modifying or repurposing a clothesline support to serve as a trellis is not a new idea, but it's certainly a good one. It's also kind of a head-slapper, as in, "Why didn't I think of that?" After all, you've got this tall, sturdy, utilitarian structure taking up space in a sunny spot that's easy to reach from the house . . . so why not grow some plants on it?

If you don't already have a clothesline support or two that you can turn into a trellis, you can build this one from scratch. The construction is easier than it looks. All of the beams and uprights are joined with special timber screws, so there's no complex or custom-fit joinery. And you can build the entire trellis in your shop or garage, then dig a couple of holes and get it set up in one go.

The basic structure of the trellis is inspired by the Torii, a traditional Japanese gateway to a shrine or other sacred place. The overhanging top beam, or lintel, is a characteristic feature for this type of structure, and in this case can be used to support hanging plants or wind chimes or simply be left as is for a clean look. The vertical spindles in the center of the trellis are made with 1½-inch-square pressure-treated stock. (You can also use cedar or redwood.) They're offset from one another in an alternating pattern for a subtle decorative effect. You can change the spacing of the spindles as needed to suit your plants, or even use a different material, such as round spindles, wire, or string.

This trellis makes a great garden feature that looks good year-round and can serve as a focal point or a divider between landscape zones. You can build just one trellis and run the clotheslines between the trellis and a fence, your house or garage, or a garden shed or other outbuilding.

A trellis such as this is not only very attractive all in it's own right, it also serves two functional roles—holding up plants and holding up laundry!

Building a Clothesline Trellis

CUTTING LIST (ONE TRELLIS)

Key	No.	Part	Dimension	Material*
A	2	Post	3½ × 3½" × 10'	4 × 4
B	1	Lintel	3½ × 3½ × 81"	4 × 4
C	2	Crossbeam	3½ × 3½ × 47"	4 × 4
D	7	Spindle	1½" × 1½ × 47½"	2 × 2
E	1	Spreader	3½" × 3½ × 8½"	4 × 4

*All lumber can be pressure-treated or all-heart cedar or redwood or other naturally rot-resistant wood.

TOOLS & MATERIALS

Miter saw
Cordless drill and bits
Nail set
Tongue-and-groove pliers or adjustable wrench
Posthole digger
6" self-drilling timber screws (24)
2" exterior finish nails
Optional: ⅜ × 2¾" galvanized or stainless-steel
  screw eyes or screw hooks
  (4, with lag-screw threads)
Gravel
Level
Eye and ear protection
Work gloves

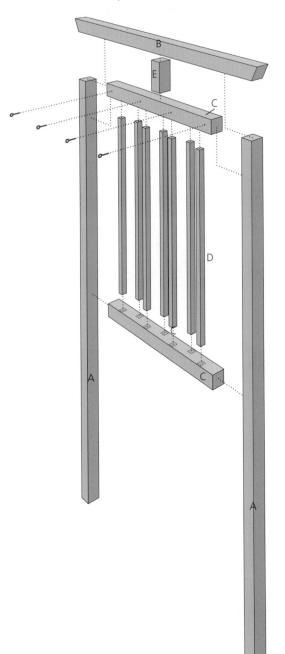

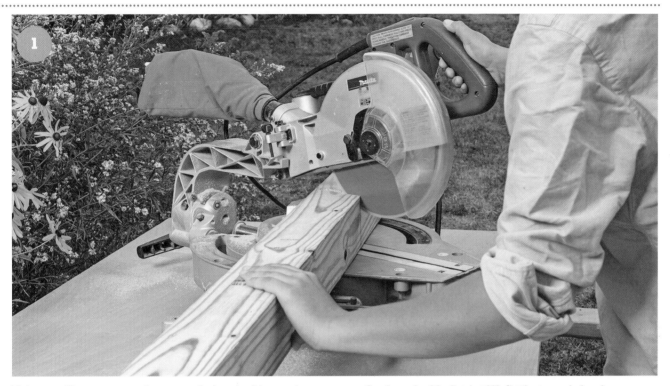

Using a miter saw or circular saw, cut both ends of the crossbeams square. Cut the ends of the lintel at 30°. Cut the top end of each 10' post to ensure a clean, square cut with no splits; the bottom end will be buried at least 3' deep, so overall length isn't critical.

Mark the inside faces of the two posts for the crossbeam locations, using a square to draw layout lines across the post faces. Mark the underside of the lintel in the same way; it is centered over the long posts, while the center post is centered on the lintel.

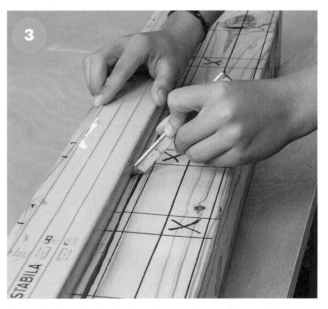

Mark the 2 × 2 spindle locations on the crossbeams. For more contrast, you can offset the locations—but remember to mark the opposing sides of the crossbeam as mirror images. Here the spindles are placed ¼" in from the edges and spaced 4⅝" apart.

(continued)

Building a Clothesline Trellis (continued)

Test-fit the frame assembly on a flat work surface. Fasten the crossbeams to the posts with two 6" self-drilling timber screws at each joint. You may need to drill pilot holes if the screws are difficult to drive. Drive the screws with either a drill or an impact driver and a hex-type nut driver or other bit (special bits often come with boxes of timber screws). If your 4 × 4s are well-dried, you can attach the lintel now; if not, save some back strain and bolt it on after the posts are upright.

Dig two holes at least 3' deep for the posts. Shovel a few inches of gravel into each hole, then tip the whole assembly in (use a helper—wet treated wood is heavy). Plumb and brace the posts with 1 × 2s.

Check that the crossbeams are level. To raise one side, simply add a little gravel under the post. Fill the postholes with alternating layers of soil and gravel. Or you can fill the holes with concrete. Check level and plumb as you fill the holes. Tamp the dirt and gravel so that the posts are firmly locked in place.

Set the lintel in position, aligning it with the marks made earlier, then fasten it in place with timber screws.

Cut the 2 × 2s to fit tightly, then fasten with predrilled 6d casing nails or a pneumatic nailer. Use two nails at each end.

Clotheslines for Climbers

Here are a few tips for planning and setting up a decorative and highly functional clothesline system:

- When the leaves are out, the trellis can create a nice a shady spot for a bench on the side opposite the clotheslines.

- A typical load of laundry needs about 35 linear feet of clothesline and weighs about 15 to 18 pounds after spinning in the washing machine. Therefore, running three or four 12' to 15' lines strikes a good balance for minimizing sag on the loaded lines while providing plenty of room for hanging.

- Clothesline materials include solid-metal wire, stranded wire or cable, plastic-coated cable, and traditional clothesline rope. Wire lines last longer and stretch less than rope, but many clothesline devotees prefer rope for its natural look and feel as well as its thickness and texture, which make it ideal for gripping fabric and clothespins.

- Pulleys allow you to hang and retrieve clothes from one standing position. Metal pulleys are strong and won't break down with UV exposure (like plastic wheels will), but make sure all metal parts are rust-resistant (stainless steel is best).

- Turnbuckles provide means for tensioning wire lines without having to restring or reclamp them. Tighten rope lines by simply retying them, or you can add a hook and a trampoline spring to maintain tension and make it easy to remove the line without untying it.

teepee trellis

19

Vertical gardening is one of the best ways to produce a large harvest in a modest amount of space. That's why trellises are so popular in gardens and why there are so many different trellis designs. The teepee style shown here is popular among homesteaders of all stripes because it's easy to build, saves space, is adaptable, and costs next to nothing.

The teepee trellis in this project is one of the simplest, including only three legs and connecting wire that creates horizontal surfaces to encourage even more plant growth. We've used widely available 1× cedar legs that you can find at any home center or lumberyard. If you're not fond of square-leg design, you can always substitute dowels or other materials (see page 131 for a rundown of different options). This design is meant to be portable, so that you can use it in different areas of the garden at different times or seasons. It also means that you can store the trellis in a shed or garage over the winter, which will prolong its life. Portability is just one aspect of the amazing adaptability in this type of trellis.

For instance, you can use longer poles to accommodate the greater vegetation of more aggressive growers. Or use more poles, creating a variety of surfaces for your vining plants to exploit. Space them wide apart, leave a large gap on one side so that the mature plants provide a shady refuge for kids, or modify any one of a number of ways that are pretty much only limited to the margins of your imagination.

No matter how you build it, your plants are sure to thrive on a teepee trellis. Although it provides a maximum of vertical growing space, it also allows for ample light penetration and airflow, ensuring against disease and keeping your plants as happy as can be.

 Vertical Plants

There are an amazing variety of plants to grow up a trellis such as this. Some are easy choices, commonly found species that you're likely to find at the local nursery or home centers. But you can also use your teepee trellis for somewhat more unusual choices.

Runner beans. These are some of the most popular choices for growing on trellises. The plants produce a wealth of tasty beans, and it's easy to put up the extra.

Peas. Peas of all kinds grow like crazy on a teepee trellis. The delicate flower and interesting shoots make them a favorite in a child's garden.

Cucumbers. A smart choice for trellis culture, cucumbers take to vertical living, and insects are less likely to attack the luscious produce on a vertical support.

Tomatoes. Surprised? Don't be. When you grow a tomato plant inside a tomato cage, you're accommodating the plant's need for support. Tie it up to a teepee trellis and the plant will flourish.

Zucchini. This is a teepee trellis favorite, one that grows like crazy with very little maintenance or care. Other squashes are equally enthusiastic about teepee trellis life, but you need to be sure that any plant producing heavier vegetables is properly supported by the teepee's framework material.

Melons. Get your melons up off the ground to do away with mushy, yellow flat spots and to keep them safe from insects. Cantaloupe, honeydew, and even watermelon can be grown vertically. Just make sure they are properly supported, and pick them as soon as they are ripe.

Peppers. Peppers of all kinds do extremely well when grown up a teepee trellis.

A teepee trellis like this is a space-conserving way to grow a large crop of runner beans or other vining vegetable. You can even grow flowering vines up one side for a showstopping garden addition.

Building a Simple Teepee Trellis

TOOLS & MATERIALS

Measuring tape
12 small screw eyes
Picture hanging wire
Pliers

Screwdrivers
Wire cutters
Garden twine or other wrapping material
Hammer

Lay the three cedar poles side by side. Measure and mark a screw eye location 24" from one end, and then 24" from the first mark. Lightly tap the point of the screw eyes into the marks using a hammer, and then screw them in using pliers or a screwdriver. Turn and repeat the process on the adjacent face of the pole.

Lean the poles together where they'll be placed in the garden, forming a loose teepee. (The screw eyes should be positioned on the inside two faces of all poles.) Tightly wrap a cord or twine around the neck of the teepee. You can also use any kind of durable, fairly weather-resistant material.

Wrap wire around one screw eye using the pliers. Stretch the wire to the same screw in an opposite leg, keeping the wire taut without pulling the legs any closer together (you can use temporary blocking if you're having trouble holding the legs in position while twisting the wire to secure it). Twist the wire around the second screw eye to secure it. Repeat with all the remaining screws. You should now be able to collapse the trellis as necessary to move it.

Teepee Trellis Alternatives

There is no lack of possibilities when it comes to the materials you can use for a teepee trellis pole. The two qualities you're looking for are durability in the face of excessive sun exposure (and all weather if you leave the trellis in position year-round), and a good strength-to-weight ratio. The stronger the material is, the heavier vegetables it can support, but if it's too light, it may need to be anchored in windy areas.

Here's a short list of common materials used in teepee trellises. Keep in mind, though, you can use anything that is non-toxic and will stay stiff under a load.

• **Bamboo.** Naturally strong, this "grass" is widely available as cut poles. Or, if you happen to have a stand of bamboo in your yard, you can cut your own bamboo poles for free. Bamboo is stiff enough to support lush vegetable growth, but it will degrade over time. In the meantime, it is one of the most attractive trellis poles you can use.

• **Branches/Found wood.** You'll have to hunt for just the right branches (hardwood preferred), but if you can locate branches that are 6 feet long or longer, and wider than 1 inch in diameter, you'll probably have yourself a very rustic-looking, sturdy teepee trellis frame. Tie branches together with rough twine to carry through the look.

• **Reclaimed plumbing.** Long, skinny plumbing pipes are nearly indestructible poles for use in a teepee trellis. You don't want to reuse any plumbing that has carried waste or toxic materials, but otherwise, it will all work. It's strong, straight,

and there are lots of ways to connect the poles for the teepee. The one big downside to a plumbing-pipe teepee trellis is the look. Most pipes are anything but handsome, but if looks are important to you, consider using copper pipes. Although the metal is attractive, you'll want to seal copper pipes because copper can be toxic to many plants.

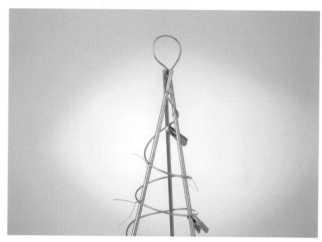

Pipes, such as these copper plumbing tubes, make a lovely teepee trellis.

• **Dowels or tool handles.** These pieces of wood are well-suited for use in a teepee trellis. They come in standard sizes that make designing the trellis easy, and they are widely available and inexpensive. You can also choose from a range of diameters, depending on how firm you need the support to be.

Anchor Ideas

Stabilizing a teepee trellis is essential in a garden where kids, pets, or barnyard animals are regular visitors (or in any windy location—once the trellis is covered in vegetation, it becomes susceptible to wind). There are lots of ways to anchor a teepee trellis, but here are three that are very effective.

Buried Poles
The simplest way to secure the posts of the teepee is to bury the ends of the posts. This is an ideal solution where you are sure that you want the trellis to be a permanent fixture. Be aware that if the pole ends are not below the frost line in your area, the poles may heave up during the winter.

Potted Poles
This is the most interesting option. Pot up your climbing vegetables in large pots (usually, 2-gallon pots or larger). Before you fill the pots with soil around the plants, hold the poles in place and then backfill around them. It becomes a simple thing to grow the plant right up the pole as it matures. You can use just about any type of pot, giving you a lot of options as far as creating an interesting look at the base of your teepee.

Staked-Wire Teepee
Have an old orphan fence post or other buried pole in your garden? Put it to good use by running guy wires off it, to stakes in the ground. This is an exceptionally strong construction. The appearance depends on the central post, because the wires quickly become invisible underneath the vegetation. As a safety measure, tie strips of plastic to the wires for visibility, at least until the plants grow over the wires.

raised bed with removable trellis

It's hard to beat PVC plumbing pipe for adding a trellis to a simple raised bed. It's inexpensive and rot-proof and goes together like pieces of a toy construction set. It's also durable, lightweight, and can stand up to just about anything the elements can throw at it.

This all-purpose trellis is made almost entirely with PVC parts and is designed to be custom-fit to your raised bed. For a bed with 2× lumber sides, you can secure the trellis uprights to the outside of the bed with metal pipe straps. If the sides of the bed are built with timbers, the trellis simply drops into holes drilled into the tops of the timbers. Of course, you can get much more creative with the configuration if you want to bump up your yield. Add another, identical trellis to the opposite end of a longer raised bed. Or add three—at either end and in the middle—of a really long bed. Making the most of vertical space with a trellis is a great way to grow a lot more vegetables in the same footprint. This can be key if you're looking to put up a lot of vegetables for over the winter.

The trellis as shown is made with 1½-inch PVC pipe and fittings. The parts are friction fit only, so they are not glued together and can easily be disassembled for off-season storage. PVC pipe and fittings are manufactured for a very tight fit; if you push the pipe all the way into the fittings, the joints won't come apart unless you want them to. Separate the joints by twisting the pipe or fitting while pulling straight out. Due to the tight fit, it doesn't help to try to wiggle it loose.

PVC pipe and fittings come in one color: stark white. You may want to paint your trellis to blend in with your garden setting, but this isn't necessary. Once it becomes covered with lush plant growth, the appearance of the pipe will be much less noticeable. Exposure to sunlight somewhat dulls PVC over time, but this doesn't significantly affect its strength.

TIP

Raised Box Trellis Options

The basic design of this bed-and-trellis combination lends itself to customization. If, for instance, you want a sturdier trellis to support much heavier plants or stand up to high winds, you can swap the PVC pipes and fittings for metal plumbing pipes and fittings. Better yet, if you want to add a rustic appeal to your unit, you can use copper pipes—just seal the copper so that it doesn't contaminate your plants. The box itself can easily be fabricated from found lumber, or lumber reclaimed from construction sites (just ask the site supervisor or foreman before you go dumpster diving). The only requirement is that the wood you reuse not be treated in any way. Otherwise, free is the best price for a self-sufficient project!

A raised bed box made from 2 × 6 lumber is used as the base for a sturdy built-in trellis made of PVC tubing. It's the ideal support for heavy climbing plants like beans or cucumbers.

Building a Raised Bed With Removable Trellis

CUTTING LIST

Key	No.	Part	Dimension	Material
A	2	Side	1½ × 5½ × 72"	2 × 6 cedar
B	2	End	1½ × 5½ × 36"	2 × 6 cedar
C	2	Upper vertical	1½ × 60"	PVC pipe
D	2	Lower vertical	1½ × 12"	PVC pipe
E	2	Crosspiece	1½ × 34"	PVC pipe

TOOLS & MATERIALS

(2) 1½" × 10' PVC pipe
(2) 1½" PVC 90° elbows
(2) 1½" PVC T-fittings
Heavy jute or hemp twine
Pipe straps for 1½" PVC (4 screw type)
Metal inside corners
Deck screws 1¼", 2½"
Tape measure
Cordless drill and bits
Hacksaw or miter saw
Sandpaper
Scissors or utility knife
Eye and ear protection
Work gloves

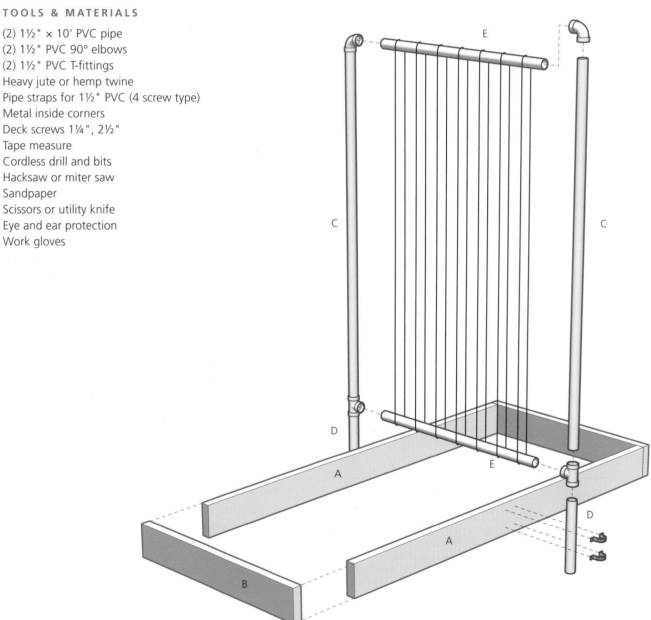

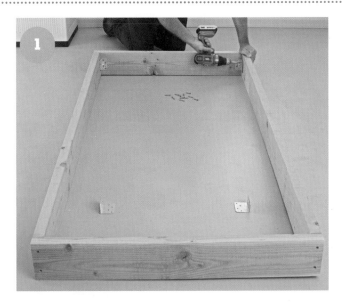

Start by assembling the raised-bed box, reinforcing the joints with metal inside corners. Add a center divider to keep the sides from spreading apart if you decide to make this project longer than 6'. Even if it is shorter, the divider is still a good precaution to help prevent warping.

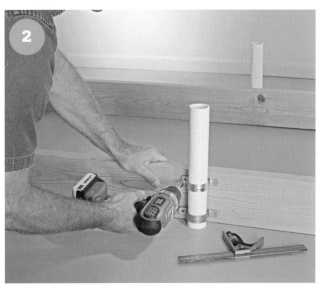

Cut 12"-long pieces of 1½" PVC tubing. Attach them to the outsides of the planter box, near the middle. Use emery paper or sandpaper to remove the burrs and smooth the cut ends of pipe. Draw a perpendicular line where the pipe will go, using a square. Strap the pieces to the outsides with two pipe straps each. Fasten one strap with two screws, but leave the other strap loose until you put the upper vertical PVC on and can check it for plumb.

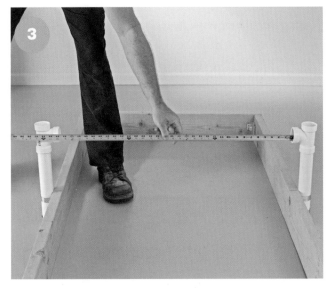

Add a T-fitting to the top end of each pipe. Measure between the hubs of the T-fittings to the insides of the sockets. Cut a piece of 1½" PVC pipe to this length and sand the cut edges smooth; this is the bottom crosspiece. Remove both Ts, fit the piece into the middle hubs of the Ts so the ends of the pipe bottom out in the fittings. Then replace the Ts.

Add the uprights and attach the top crosspiece with elbows. Ensure the pipes are plumb, then secure the bottom straps. Move the planter into your yard or garden, line it with a thick layer of old newspaper or landscape fabric, and fill it with planting medium. Tie jute or hemp twine between the crosspieces so that climbing plants have something to grab onto. When winter comes, you can disassemble the PVC and store it away until spring.

pallet planter

Pallets are so abundant in this country that they're often just left by the curb for people to take as firewood. Despite this lack of perceived value, they can actually be quite useful to the thrift-minded self-sufficient homeowner. They can be perfect for a wide variety of recycled projects like compost bins, furniture, fuel and, yes, planters.

Because pallets have to support thousands of pounds, they're generally made from tough hardwoods without big knots. When the wood is cleaned up and sanded, it can look surprisingly attractive. Pallet wood may not last as long outdoors as cedar or treated wood (unless you find a pallet made from white oak), but since it's free you can just replace it when it rots. In the meantime, it looks great.

There are many different ways to turn a pallet into a planter, but don't count on just pulling out all the nails and reusing all the pieces. It can be done, but hardwood grips nails a lot tighter than the softwoods used for construction lumber, and the wood slats often crack before the nail pulls out. (It's wise to always grab a few extra pallets for any project, precisely for this reason.) If you have to remove a few pieces, lever carefully under the wood slats, and use a reciprocating saw outfitted with a metal-cutting blade to cut nails that are hard to pull. However, the easiest way to use pallets is as is—as we've done in this project. We designed this one to be vertical, but you can also lay it flat and stack a few underneath so that the plants are at a comfortable height, or combine several in a stairstep design.

Look for pallets in industrial and commercial areas. If you're lucky, you'll see a big pile with a "Free" sign on them, but you can also find them poking out of dumpsters or just piled up in a parking lot. If in doubt, always ask if you can take them. Most will be dirty and have a cracked or missing slat, but if you grab an extra you can use it for parts.

TIP

Pallet Shopping

There is actually a good deal of information stamped on every pallet. Obviously, you should avoid those that look heavily oil-stained, or smell of chemicals. But beyond that, you can find the label on the pallet and use it to tell you something about the wood itself. The stamp will usually show the logo of the company who produced it. It will also list what country it was made in, and—most importantly—it will list a code that usually starts with "DB" (de-backed) and includes two letters that indicate how the wood was treated. "HT" means the wood was heat treated and will be fine for use in contact with soil. "MB" means the wood was treated with the pesticide methyl bromide, and should probably not be used for food crops.

Upcycling is a key part of any self-sufficient household, and pallets are some of the handiest candidates for the treatment. For this vertical planter project, the pallet is left intact and converted for use in a smaller area such as a patio, deck, or balcony. All part of turning every corner of the yard into a productive part of the whole.

Making a Pallet Planter

After brushing off the dirt and renailing any loose boards, use some 80-grit sandpaper to clean up the areas that will be visible. Also round over rough, splintery edges. Paint or finish the outside of the pallet, if desired, but don't finish the inside if you're planting edibles.

Making the Planter

The process used to turn this pallet into a planter is a simple one and can be done by anyone with even very basic DIY skills. Decide which end of the pallet will be up, then cover openings at the back to keep dirt from falling out through the openings.

Gaps on the sides will be covered with a strip of metal flashing, an extra slat or a piece of wood cut to size. The back is covered with rubber pond liner or a double layer of black 6-mil poly (but make sure none of it is exposed to the sun, or it will decay).

Finally, nail a doubled-over strip of aluminum screen mesh across the bottom of the pallet to keep the dirt in, and then fill with topsoil. Tamp it down with a long stick to make sure the pallet fills up. Dirt will fall out the front at first, but will settle in at an angle behind each opening.

You can simply lean the planter against a wall in a sunny area if you prefer. However, you can also mount the planter on a southern or eastern-facing side of a garage, house, or outbuilding. If you prefer to put the planter out in the garden or in a sunny spot in the yard, you can screw stakes to each side or front and back, and then secure it in the ground so that it doesn't fall over when bumped or on a windy day.

If space is tight, you can even wire it to a balcony or deck railing. Just be sure the railing is solid because when you water the plants and soil in the planter, it can become fairly heavy.

Pallet Planter Options

You can make individual planters using the same method described here, but by cutting the pallet into sections. You can also decorate the pallet to better suit the style of the yard. As long as you don't get any paint inside, you can paint the outside in different colors or even stain the wood if you prefer.

Planting Your Pallet Planter

As you examine your completed pallet planter, one question immediately comes to mind: won't all the dirt fall out the front? Well, if you did not plant any plants in it the answer is yes, it will. However, you are relying on the plant roots and the 2 × 2 shelves you installed to hold things together. To add plants, lay the planter flat on its back and plan to keep it that way for a couple of weeks. Pack the gaps full of potting soil (potting soil is pre-fertilized, unlike topsoil) and then pack in as many seedlings as you can fit. Water the plants for a couple of weeks so the roots can establish. Then, tip the planter up against the wall in position. A little soil may trickle out initially, but you should find that everything holds together nicely.

Best Pallet Plants

Not every vegetable or fruit will be at home in a pallet planter, but it can serve as the ideal location for several garden favorites.

Strawberries. Just as in a strawberry pot, these plants are at home with the shallow soil in the planter "pockets." The plants are easy to work with and, once settled in the planter, will produce a crop for several years.

Leaf lettuces. The plants will spread out in the pockets and, if kept watered and reasonably cool (you may need to shade the planter during the hottest part of the day), they will thrive in the planter.

Herbs. A pallet planter is idea for an herb garden. Because you only need a small amount of each herb, the planter can support an entire kitchen herb garden. It can also be set against a wall right outside a kitchen door, making the herbs incredibly convenient for harvesting.

TOOLS & MATERIALS

Cedar 2 × 2s, cut to fit
Cordless drill and bits
Screwdriver or pry bar
2" deck screws
Staple gun and staples
Pond liner or 6-mil poly
Scrap wood
Aluminum screen mesh
Eye and ear protection
Work gloves

Cut the 2 × 2 shelves to the length for the cavities in your pallet and fit them in on the lower edge of each slat. Predrill and toenail at each end to hold them in place, then drive an additional screw into the cedar from the front. Pallet wood is hard, so drill pilot holes for all screws. **Note:** Most pallets have a good face and a bad face. Be sure that you are creating your planter so the face with the nicer decking will point away from the wall you are installing it against.

Stiffen the pallet and provide a surface for attaching the liner. Fill the gaps on the back side with wood scraps of roughly the same thickness as the back boards.

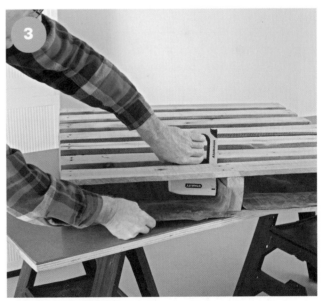

Measure and cut pond liner or poly sheeting for the back. Fold the liner onto the sides and staple it in place on the sides and back. You can use heavy black poly sheeting to make the liner, but for a more durable material use roll rubber. Staple aluminum screen mesh across the bottom of the pallet.

raised beds

Raised garden beds offer several advantages over planting at ground level. When segregated, soil can be amended in a more targeted way to support high density plantings. Also, in raised garden beds, soil doesn't suffer compaction from foot traffic or machinery, so plant roots are free to spread and breathe more easily. Vegetables planted at high densities in raised beds are placed far enough apart to avoid overcrowding, but close enough to shade and choke out weeds. In raised beds, you can also water plants easily with soaker hoses, which deliver water to soil and roots rather than spraying leaves and inviting disease. And if your plants develop a fungus or another disorder, it is easier to treat and less likely to migrate to other plants in a raised bed situation.

TIP

Bed Positions

If you're planting low-growing crops, position the bed with a north-south orientation, so both sides of the bed will be exposed to direct sunlight. For taller crops, position the bed east-west.

Raised garden beds can be built in a wide variety of shapes and sizes, and can easily be customized to fit the space you have available on your property. Just make sure you can reach the center easily. If you can only access your raised bed from one side, it's best to build it no wider than 3 feet. Beds that you can access from both sides can be as wide as 6 feet, as long as you can reach the center. You can build your raised bed as long as you'd like.

Raised garden beds can be built from a wide variety of materials: 2× lumber, 4 × 4 posts, salvaged timbers, even scrap metals and other recycled goods. Make sure any lumber you choose (either new or salvaged) hasn't been treated with creosote, pentachlorophenol, or chromated copper arsenic (CCA). Lumber treated with newer,

Raised garden beds make great vegetable gardens—they're easy to weed, simple to water, and the soil quality is easier to control, ensuring that your vegetable plants yield bountiful fresh produce. Your garden beds can be built at any height up to waist-level. It's best not to build them much taller than that, however, to make sure you can reach the center of your bed.

Vegetable Plant Compatibility Chart

Vegetable	Loves	Incompatible with	Planting Season
Asparagus	Tomatoes, parsley, basil		Early spring
Beans (bush)	Beets, carrots, cucumbers, potatoes	Fennel, garlic, onions	Spring
Cabbage & broccoli	Beets, celery, corn, dill, onions, oregano, sage	Fennel, pole beans, strawberries, tomatoes	Spring
Cantaloupe	Corn	Potatoes	Early summer
Carrots	Chives, leaf lettuce, onion, parsley, peas, rosemary, sage, tomatoes	Dill	Early spring
Celery	Beans, cabbages, cauliflower, leeks, tomatoes		Early summer
Corn	Beans, cucumbers, peas, potatoes, pumpkins, squash		Spring
Cucumbers	Beans, cabbages, corn, peas, radishes	Aromatic herbs, potatoes	Early summer
Eggplant	Beans	Potatoes	Spring
Lettuce	Carrots, cucumbers, onions, radishes, strawberries		Early spring
Onions & garlic	Beets, broccoli, cabbages, eggplant, lettuce, strawberries, tomatoes	Peas, beans	Early spring
Peas	Beans, carrots, corn, cucumbers, radishes, turnips	Chives, garlic, onions	Early spring
Potatoes	Beans, cabbage, corn, eggplant, peas	Cucumber, tomatoes, raspberries	Early spring
Pumpkins	Corn	Potatoes	Early summer
Radishes	Beans, beets, carrots, cucumbers, lettuce, peas, spinach, tomatoes		Early spring
Squash	Radishes	Potatoes	Early summer
Tomatoes	Asparagus, basil, carrots, chive, garlic, onions, parsley	Cabbages, fennel, potatoes	Dependent on the variety
Turnips	Beans, peas		Early spring

non-arsenate chemicals at higher saturation levels is rated for ground contact and is also a safe choice for bed frames. Rot-resistant redwood and cedar are good choices that will stand the test of time. Other softwoods, including pine, tamarack, and cypress, will also work, but can be subject to rot and may need to be replaced after a few years.

○ Companion Planting

The old adage is true—some vegetables do actually get along "like peas and carrots." Some species of vegetables are natural partners that benefit from each other when planted close. On the other hand, some combinations are troublesome, and one plant will inhibit the growth of another. You can plant these antagonists in the same garden—even in the same raised garden bed—just don't place them side by side. Use the table on page 142 to help you plan out your raised garden beds to ensure that your plants grow healthy, strong, and bear plentiful fruit.

○ Start with Healthy Soil

The success or failure of any gardening effort generally lies beneath the surface. Soil is the support system for all plants—it provides a balanced meal of the nutrients that plants' roots need to grow deep and strong. If you plan to fill your raised beds with soil from your property, it's a good idea to have the soil tested first to assess its quality. Take a sample of your soil and submit it to a local agricultural extension service—a basic lab test will cost you between $15 and $25 and will give you detailed information about the nutrients available in your property's soil. Mixing soil from your property with compost, potting soil, or other additives is a smart and inexpensive way to improve its quality. After you've filled your beds with soil, add a 3-inch layer of mulch to the top to lock in moisture and keep your good soil from blowing away in strong winds. Lawn clippings, wood, or bark chips, hay and straw, leaves, compost, and shredded newspaper all work well as mulch materials.

Watering Raised Beds

When the soil inside the planting bed pulls away from the edges of the bed, it's time to water. The best time of day to water is in the late afternoon or early to midmorning. Avoid watering in midday, when the sun is hottest and water will quickly evaporate, or near sundown or at night, when too much moisture in the soil can cause mold and fungus to grow. (For another watering method, see page 94.)

Sprinklers with high, arching spray patterns are afflicted by excess water evaporation, but if you choose a small, controllable sprinkler with a water pattern that stays low to the ground you can deliver water to your raised bed with minimal loss.

How to Build a Raised Bed

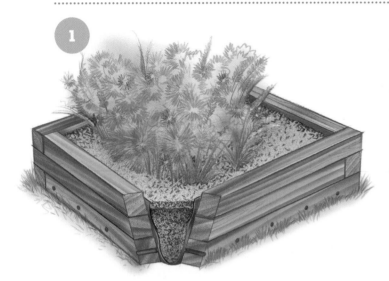

This basic but very sturdy raised bed is made with 4 × 4 landscape timbers stacked with their ends staggered in classic log-cabin style. The corners are pinned together with 6" galvanized spikes (or, you can use timber screws). It is lined with landscape fabric and includes several weep holes in the bottom course for drainage. Consider adding a 2 × 8 ledge on the top row. Corner finials improve the appearance and provide hose guides to protect the plants in the bed.

Create an outline around your garden bed by tying mason's string to the stakes. Use a shovel to remove the grass inside the outline, then dig a 4"-wide trench for the first row of timbers around the perimeter. Lay the bottom course of timbers in the trenches. Where possible, add or remove soil as needed to bring the timbers to level—a level bed frame always looks better than a sloped one. If you do have a significant slope to address, terrace the beds.

Add the second layer, staggering the joints. Drill pilot holes at the corners and drive 6" galvanized spikes (or 6 to 8" timber screws) through the holes—use at least two per joint. Continue to build up layers in this fashion, until your bed reaches the desired height.

Line the bed with landscape fabric to contain the soil and help keep weeds out of the bed. Tack the fabric to the lower part of the top course with roofing nails. Some gardeners recommend drilling 1"-dia. weep holes in the bottom timber course at 2' intervals. Fill with a blend of soil, peat moss, and fertilizer (if desired) to within 2 or 3" of the top.

How to Build a Raised Bed from a Kit

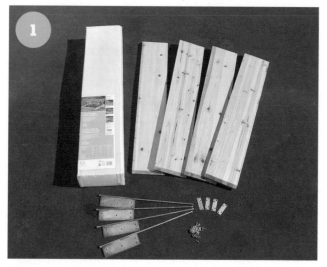

Raised garden bed kits come in many styles. Some have modular plastic or composite panels that fit together with grooves or with hardware. Others feature wood panels and metal corner hardware. Most kits can be stacked to increase bed height.

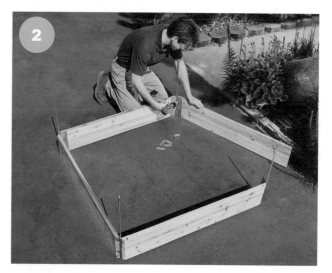

On a flat surface, assemble the panels and corner brackets (or hinge brackets) using the included hardware. Follow the kit instructions, making sure all corners are square.

Set the box down, experimenting with exact positioning until you find just the spot and angle you like. Be sure to observe the sun over an entire day when choosing the sunniest spot you can for growing vegetables. Cut around the edges of the planting bed box with a square-nose spade, move the box, and then slice off the sod in the bed area.

Set the bed box onto the installation site and check it for level. Add or remove soil as needed until it is level. Stake the box to the ground with the provided hardware. Add additional box kits on top of or next to the first box. Follow the manufacturer's instructions for connecting the modular units. Line the bed or beds with landscape fabric and fill with soil to within 2" or so of the top.

container gardening

23

If you're short on yard space and a raised bed doesn't work with your landscaping, you can still plant a productive vegetable garden by utilizing sunny spots on balconies, stoops, porches, windowsills, and other out-of-the-way spaces. Many plants flourish in small planters or other portable containers, and with a few well-managed containers, you can yield a sizable bounty for a couple or small family. In some ways, container gardening is easier than having a full-blown vegetable garden. There are fewer weeds (if any), pests are less problematic, and diseases are easier to avoid. You also need fewer tools and plants can easily be moved to accommodate temperature fluctuations and light/shade patterns. If you're a beginner gardener just wanting to try your hand at growing your own food, container gardening could be a great solution for you.

The key to successful container gardening is water. Because of their limited size, even the largest containers simply cannot hold the amount of water the plants need to thrive without watering once every day—extremely thirsty plants may even need to be watered

Fresh Soil

Do not reuse soil if you're growing tomatoes or other plants that are susceptible to blight and fungus. After the growing season is concluded, collect the soil from sensitive plant containers and disperse it into hardy planting beds.

If you've never tried gardening, container gardening is a good way to start.

What to Grow

Most plants that grow well in your garden will also thrive in containers. Root vegetables are perhaps the only exception to this rule. Keep in mind that the larger the plant, the larger the container you'll need. Generally, the plant should not be more than twice the height of the pot or 1½ times as wide. Use the guidelines below as a rough guide.

Containers 4 to 6" deep: Mustard greens, radishes, and spinach can all grow in shallow containers.

Containers at least 8" deep: Corn (container must be at least 21" wide, however, and house at least three plants to assure pollination), kale, lettuce.

Containers at least 12" deep: Beans, beets, brussel sprouts, cabbage (should also be pretty wide), carrots, chard, collards, kohlrabi, onion, peas, turnips, zucchini.

Containers at least 16" deep: Cucumber, eggplant, peppers.

Containers at least 20" deep: Broccoli, bok choy, Chinese cabbage.

More than 24" deep: Squash, tomatoes.

more frequently than that. Vegetable plants are especially thirsty, but herbs and fruit trees or bushes need careful watering as well. Mature tomato plants may need as much as a gallon of water every day to grow those juicy, delicious fruits. If you bury your finger in the pot and you feel any dry soil—even 2" down—it's time to water.

Soil in container gardens is important too. A reliable rule of thumb is to use a 50/50 mix of potting soil to compost. Soil from your property typically won't hold water as well as potting soil on its own, so it's best not to use it in container gardens without adding a significant amount of organic fertilizer, which you can buy at the garden center or mix yourself. The following pages provide guidelines about the types of containers you can use and step-by-step instructions to help you build your own planter boxes.

Container Types and Recommendations

As a container gardener, you'll quickly discover that the universe of usable containers is infinitely larger than the plain clay flowerpot. Essentially, any sturdy, watertight container will do. Large containers like wine barrels or old wash tubs and smaller containers like an ice cream pail or 5-gallon bucket can all be good for different kinds of plants. Large wooden troughs and DIY planter boxes can be customized to your garden (and are fun to make too). When building your own planters, it's a good idea to line the inside with landscape fabric before adding potting soil to protect the wood from rot and to make it easier to empty out soil after a season.

Always make sure that the container you choose did not previously hold any kind of chemical and, if it does not already have them, drill drainage holes near the bottom of the container before filling with soil. If you'll be using large containers, it's usually a good idea to place them on a platform fitted with casters before filling them with potting soil.

Self-watering pots make container gardening less of a drain on your time. These containers are, essentially, a flowerpot set just above a reservoir of water. With this type of container, the soil above will wick up moisture from the reservoir as it needs it—keeping the soil consistently moist throughout and eliminating the possibility of over-watering. With a self-watering container, you may only need to add water every three to four days, and your plants will likely be less stressed. Ideally, your plants will therefore provide a more sizable crop at the end of the season. (See page 94 for a larger version of a self-watering container.)

Spinach, leaf lettuce, and a few shallots co-exist in this self-watering planter. Self-watering containers have a water reservoir below to keep the soil in the pot moist. All you have to do is keep the reservoir full, and rainfall may even take care of this for you.

Many varieties of vegetables and fruits can be grown in pots, planters, and other vessels—just make sure to select a container that is the right size for your plant and always add ample drainage holes if they're not already present.

○ Planter Boxes

Decorating a garden is much like decorating a room in your home—it's nice to have pieces that are adaptable enough that you can move them around occasionally and create a completely new look. After all, most of us can't buy new furniture every time we get tired of the way our living rooms look. And we can't build or buy new garden furnishings every time we want to rearrange the garden.

That's one of the reasons this trio of planter boxes works so well. In addition to being handsome—especially when flowers are bursting out of them—they're incredibly adaptable. You can follow these plans to build a terrific trio of planter boxes that will go well with each other and will complement most gardens, patios, and decks. Or you can tailor the plans to suit your needs. For instance, you may want three boxes that are exactly the same size. Or you might want to build several more and use them as a border that encloses a patio or frames a terraced area.

Whatever the dimensions of the boxes, the basic construction steps are the same. If you decide to alter the designs, take a little time to figure out the new dimensions and sketch plans. Then devise a new cutting list and do some planning so you can make efficient use of materials. To save cutting time, clamp together parts that are the same size and shape and cut them as a group (called gang cutting).

When your planter boxes have worn out their welcome in one spot, you can easily move them to another, perhaps with a fresh coat of stain, and add new plantings. You can even use the taller boxes to showcase outdoor relief sculptures—a kind of alfresco sculpture gallery.

Whether you build only one or all three, these handy cedar planters are small enough to move around your gardens and inside your greenhouse or garden shed.

Building Planter Boxes

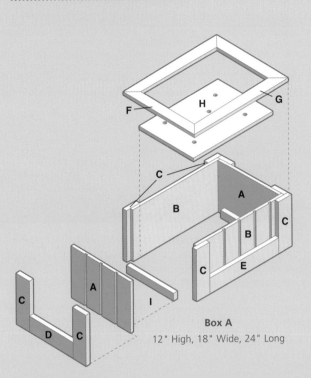

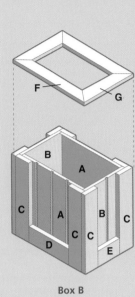

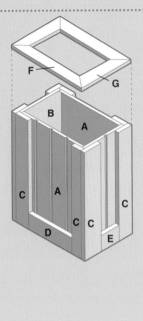

Box A
12" High, 18" Wide, 24" Long

Box B
18" High, 18" Wide, 12" Long

Box C
24" High, 18" Wide, 12" Long

CUTTING LIST

Key	No.	Part	Box A Dimension	Box B Dimension	Box C Dimension	Material
A	2	End panel	⅝ × 15 × 11⅛"	⅝ × 15 × 17⅛"	⅝ × 15 × 23⅛"	Siding
B	2	Side panel	⅝ × 22¼ × 11⅛"	⅝ × 10¼ × 17⅛"	⅝ × 10¼ × 23⅛"	Siding
C	8	Corner trim	⅞ × 3½ × 11⅛"	⅞ × 3½ × 17⅛"	⅞ × 3½ × 23⅛"	Cedar
D	2	Bottom trim	⅞ × 3½ × 9¼"	⅞ × 3½ × 9¼"	⅞ × 3½ × 9¼"	Cedar
E	2	Bottom trim	⅞ × 3½ × 17"	⅞ × 3½ × 5"	⅞ × 3½ × 5"	Cedar
F	2	Top cap	⅞ × 1½ × 18"	⅞ × 1½ × 18"	⅞ × 1½ × 18"	Cedar
G	2	Top cap	⅞ × 1½ × 24"	⅞ × 1½ × 12"	⅞ × 1½ × 12"	Cedar
H	1	Bottom panel	¾ × 14½ × 19½"	¾ × 14½ × 8½"	¾ × 14½ × 8½"	Plywood
I	2	Cleat	⅞ × 1½ × 12"	⅞ × 1½ × 12"	⅞ × 1½ × 12"	Cedar

Note: Measurements reflect the actual size of dimension lumber.

TOOLS & MATERIALS

Tape measure
Circular saw
Straightedge
Drill
Finishing sander

Miter box and backsaw
(3) 8' cedar 1 × 2s
(6) 8' cedar 1 × 4s
4 × 8' sheet of ⅝" T1-11 siding
2 × 4' piece ¾" CDX plywood

1¼" galvanized deck screws
1⅝" galvanized deck screws
6d galvanized finish nails
Exterior wood stain
Paintbrush
Eye and ear protection

Landscape fabric
Gravel
Potting soil or compost
Locking wheels or casters with brass
  or plastic housings
Work gloves

How to Build Planter Boxes

Cut all the wood parts to size according to the Cutting List on page 151. Use a circular saw and a straightedge cutting guide to rip siding panels (if you have access to a tablesaw, use that instead). You can make all three sizes, or any combination you choose.

Assemble the box frame. Place the end panel face down and butt it against a side panel. Mark the locations of several fasteners on the side panel. Drill counterbored ³⁄₃₂" pilot holes in the side panel at the marked locations and fasten the side panel to the end panel with 1⅝" deck screws. Fasten the opposite side panel the same way. Attach the other end panel with deck screws.

Attach the corner trim. Position one piece of corner trim flush to the corner edge and fasten to the panels with 1⅝" galvanized deck screws driven into the trim from the inside of the box. Place the second piece of trim flush to the edge of the first piece, creating a square butt joint. Attach to the panel with 1⅝" galvanized deck screws. For extra support, end nail the two trim pieces together at the corner with galvanized finish nails.

Attach the bottom trim. Fasten the bottom trim to the end and side panels, between the corner trim pieces and flush with the bottom of the box. Drive 1¼" deck screws through the panels from the inside to fasten the trim pieces to the box.

Attach the cap pieces. Cut 45° miters at both ends of one cap piece using a miter box and backsaw or a power miter saw. Tack this piece to the top end of the box, with the outside edges flush with the outer edges of the corner trim. Miter both ends of each piece and tack to the box to make a square corner with the previously installed piece. Once all caps are tacked in position and the miters are closed cleanly, attach the cap pieces using 6d galvanized finish nails.

Install the cleats to hold the box bottom in place. Screw to the inside of the end panels with 1⅝" deck screws. If your planter is extremely tall, fasten the cleats higher on the panels so you won't need as much soil to fill the box. If doing so, add cleats on the side panels as well for extra support.

Finish and install the bottom. Cut the bottom panel to size from ¾"-thick exterior-rated plywood. Drill several 1"-dia. drainage holes in the panel and set it onto the cleats. The bottom panel does not need to be fastened in place, but for extra strength, nail it to the cleats and box sides with galvanized finish nails.

Finish the box or boxes with wood sealer-preservative. When the finish has dried, line the planter box with landscape fabric, stapling it at the top of the box. Trim off fabric at least a couple of inches below the top of the box. Add a 2"-layer of gravel or stones, then fill with a 50/50 mix of potting soil and compost. **TIP:** Add wheels or casters to your planter boxes before filling them with soil. Be sure to use locking wheels or casters with brass or plastic housings.

○ Building a Strawberry Barrel

Container gardens aren't just for vegetables—fruit trees and berry bushes also thrive in a potted garden environment and can produce enough fruit for a family to enjoy throughout the summer. Strawberries, which typically grow in long rows or patches in the ground, can also be grown in a converted barrel. Two can be enough space to grow the equivalent of 25 feet of strawberry plants. For this project, make sure you choose everbearing strawberry varieties, and cut off runners for transplanting when they appear. Insulate your barrels with hay or straw during the winter, and you can enjoy a strawberry crop for several years. To keep your barrel going, start fresh every four or five years with new plants and new soil.

TOOLS & MATERIALS

Large, clean barrel (55-gallon plastic or wood)

Pry bar or jigsaw

3"-dia. hole saw

Drill

4"-dia. PVC pipe

Window screen or hardware cloth

Gravel

Potting soil mix

Strawberry plants

Eye and ear protection

Work gloves

A strawberry barrel planter can grow the equivalent of 25' of strawberry plants; choose everbearing varieties for best results.

How to Build a Strawberry Barrel

Prepare the barrel. If your barrel has a lid or closed top, remove it with a pry bar. If your barrel does not have a lid, cut a large opening in the top with a jigsaw. Beginning about 1' above the ground, use a hole saw to cut 3"-dia. planting holes around the barrel, about 10" apart. Stagger the holes diagonally in each row and space the rows about 10" apart. Leave at least 12" above the top row of holes. Flip the barrel over and drill about a half dozen ½"-dia. drainage holes in the bottom.

Prepare the watering pipe. Cut a section of 4"-dia. PVC pipe to fit inside the barrel from top to bottom. Punch or drill ¾"-dia. holes in the pipe every 4 to 6", all the way around. Cut a section of window screen or hardware cloth to fit inside the bottom of the barrel and place it inside. Cover the screen with 2" of gravel or small rocks.

Begin to fill the barrel with soil. Have a friend hold the watering pipe in the center of the barrel and fill the pipe with coarse gravel. Then, begin to add soil to the bottom of the barrel, packing it firmly around the watering pipe in the bottom with a piece of scrap lumber. Add water to help the soil settle. Continuing adding soil until you reach the bottom of your first row of planting holes.

Add soil and plants. Carefully insert your strawberry plants into the holes, spreading the roots into a fan shape. Add soil on top of the roots and lightly water. Continue to add soil and plants, packing soil gently and watering after each planting, until you reach the top of the barrel. Do not cover the watering pipe. Plant additional strawberries on top of the barrel. Insert a hose into the watering pipe and run water for several minutes to give the barrel a good soaking.

cold frame

24

An inexpensive foray into greenhouse gardening, a cold frame is practical for starting plants six to eight weeks earlier in the growing season and for hardening off seedlings. Basically, a cold frame is a box set on the ground and topped with glass or plastic. Although mechanized models with thermostatically controlled atmospheres and sash that automatically open and close are available, you can easily build a basic cold frame yourself from materials you probably already have around the house.

The back of the frame should be about twice as tall as the front so the lid slopes to a favorable angle for capturing sunrays. Build the frame tall enough to accommodate the maximum height of the plants before they are removed. The frame can be made of brick, block, plastic, wood, or just about any material you have on hand. It should be built to keep drafts out and soil in.

If the frame is permanently sited, position it facing south to receive maximum light during winter and spring and to offer protection from wind. Partially burying it takes advantage of the insulation from the earth, but it also can cause water to collect and the direct soil contact will shorten the lifespan of the wood frame parts. Locating your frame near a wall, rock, or building adds additional insulation and protection from the elements. **TIP:** The ideal temperature inside is 65 to 75 degrees Fahrenheit during the day and 55 to 65 degrees at night. Keep an inexpensive thermometer in a shaded

Starting plants early in a cold frame is a great way to get a head start on the growing season. A cold frame is also a great place for hardening off delicate seedlings to prepare them for transplanting.

spot inside the frame for quick reference. A bright spring day can heat a cold frame to as warm as 100 degrees, so prop up or remove the cover as necessary to prevent overheating. And remember, the more you vent, the more you should water. On cold nights, especially when frost is predicted, cover the box with burlap, old quilts, or leaves to keep it warm inside.

A cold frame should only be used during the early, cooler days of the growing season when delicate seedlings need that extra protection, and for late-season frost protection. Once the warmer weather arrives and the plants are established, remove and relocate the cold frame. Ongoing usage will overheat and kill the plants. And while the clear acrylic lid on this cold frame is desirable because it is safer to work with and use than glass, too much heat buildup can cause the acrylic to warp.

Building a Cold Frame

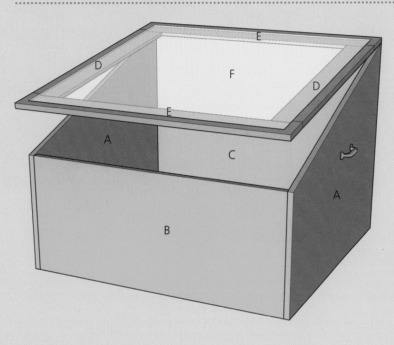

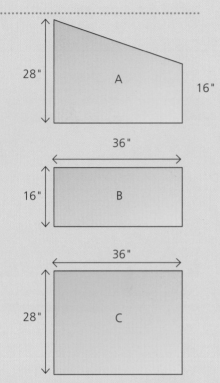

Plywood Cutting Diagram

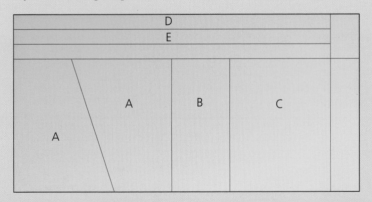

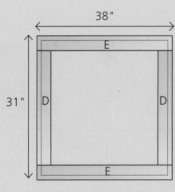

CUTTING LIST

Key	Part	No.	Size	Material
A	Side	2	¾ × 28 × 36"	Ext. Plywood
B	Front	1	¾ × 16 × 36"	Ext. Plywood
C	Back	1	¾ × 28 × 36"	Ext. Plywood
D	Lid frame	2	¾ × 4 × 31"	Ext. Plywood
E	Lid frame	2	¾ × 4 × 38"	Ext. Plywood
F	Cover	1	⅛ × 37 × 38"	Plexiglas

TOOLS & MATERIALS

(2) 3 × 3" butt hinges (ext.)

(2) 4" utility handles

(4) Corner L-brackets (¾ × 2½")

(1) ¾" × 4 × 8' Plywood (Ext.)

⅛ × 37 × 38" clear Plexiglas

Exterior caulk/adhesive

Exterior wood glue

Eye and ear protection

Exterior paint in darker color

2" deck screws

#8 × ¾" wood screws

Circular saw

Drill/driver

Pipe or bar clamps

Straightedge cutting guide

Work gloves

How to Build a Plywood Cold Frame

Cut the parts. This project, as dimensioned, is designed to be made entirely from a single 4 × 8 sheet of plywood. Start by cutting the plywood lengthwise to make a 36"-wide piece. **TIP:** Remove material in 4" wide strips and use the strips to make the lid frame parts and any other trim you may want to add.

Trim the parts to size with a circular saw or jigsaw and cutting guide. Mark the cutting lines first (See Diagram, page 159).

Assemble the front, back, and side panels into a square box. Glue the joints and clamp them together with pipe or bar clamps. Adjust until the corners are square.

Reinforce the joints with 2" deck screws driven through countersunk pilot holes. Drive a screw every 4 to 6" along each joint.

Make the lid frame. Cut the 4"-wide strips of ¾" plywood reserved from step 1 into frame parts. Assemble the frame parts into a square 38 × 39" frame. There are many ways to join the parts so they create a flat frame. Because the Plexiglas cover will give the lid some rigidity, simply gluing the joints and reinforcing with an L-bracket at each inside corner is adequate structurally.

Paint the box and the frame with exterior paint, preferably in an enamel finish. A darker color will hold more solar heat.

Lay thick beads of clear exterior adhesive/caulk onto the tops of the frames and then seat the Plexiglas cover into the adhesive. Clean up squeeze-out right away. Once the adhesive has set, attach the lid with butt hinges and attach the handles to the sides.

Move the cold frame to the site. Clear and level the ground where it will set. Some gardeners like to excavate the site slightly.

jumbo cold frame

A cold frame of any size works on the same principle as a greenhouse, capturing sunlight and heat while protecting plants from cold winds and frost. But when your planting needs outgrow a basic backyard cold frame with a window-sash roof, it makes sense to look to the greenhouse for more comprehensive design inspiration. This jumbo version offers over 17 square feet of planting area and combines the convenience of a cold frame with the full sun exposure of a greenhouse. Plus, there's ample height under the cold frame's canopy for growing taller plants.

The canopy pivots on hinges and can be propped all the way up or partially opened to several different positions for ventilating the interior to control temperature. The hinges can be separated just like door hinges (in fact, they are door hinges), so you can remove the canopy for the off season, if desired. Clear polycarbonate roofing panels make the canopy lightweight yet durable, while admitting up to 90 percent of the sun's UV rays (depending on the panels you choose).

The base of the cold frame is a simple rectangle made with 2 × 6 lumber. You can pick it up and set it over an existing bed of plantings, or give it a permanent home, perhaps including a foundation of bricks or patio pavers to protect the wood from ground moisture. For additional frost protection and richer soil for your seedlings, dig down a foot or so inside the cold frame and work in a thick layer of mulch. Because all sides of the canopy have clear glazing, you don't have to worry about orienting the cold frame toward the sun, as virtually all of the interior space is equally exposed to light.

A cold frame can extend the growing season in your garden to almost—or truly—year-round. Use an oversized cold frame like the one in this project and there may be no need to put up vegetables in the fall, because you'll have all the fresh produce you can handle.

Keeping Your Cold Frame Cool

Cold frames often can work too well, capturing and retaining so much heat that it becomes too hot for the plants, even during very cold weather. Adding an outdoor thermometer with a remote sensor (wired or wireless) lets you monitor the temp inside the cold frame without having to lift the canopy. Make sure the thermometer is rated for sub-freezing temperatures, since it will be exposed to the elements. Secure the sensor inside the frame as directed by the manufacturer. Mount the readout unit to the outside of the cold frame base. As an alternative, you can use a wireless system to send a readout to a thermometer inside the house. As a general guideline, the interior temperature of a cold frame should be no higher than 75 degrees Fahrenheit for summer plants and 65 degrees Fahrenheit or lower for spring and fall plants. But check the recommendations for your specific plantings.

Building a Jumbo Cold Frame

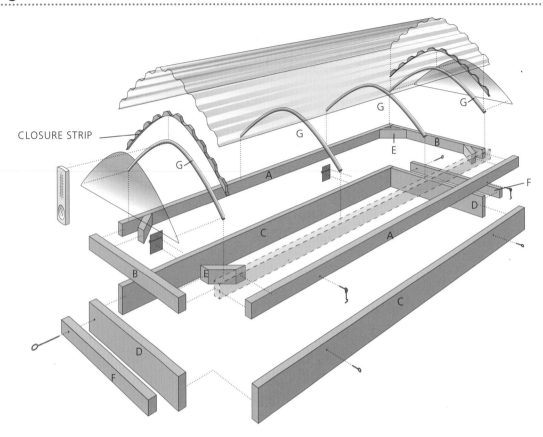

CLOSURE STRIP

CUTTING LIST

Key	No.	Part	Dimension	Material
A	2	Frame side	1½ × 2½ × 94"	2 × 3
B	2	Frame end	1½ × 2½ × 30"	2 × 3
C	2	Base side	1½ × 5½ × 94"	2 × 6
D	2	Base end	1½ × 5½ × 30"	2 × 6
E	4	Frame brace	1½ × 2½ × 8"	2 × 3
F	2	Prop stick	¾ × 1½ × 30"	1 × 2
G	4	Rib	½ × ½ × 37"	½ PVC tubing

TOOLS & MATERIALS

Circular saw or miter saw
Cordless drill and bits
Hacksaw
Deck screws 2", 2½", 3"
(5) ½" × 10' thin wall PVC pipes (the flexible type
 used for lawn irrigation, not schedule 40 type)
(2) 25 × 96" corrugated polycarbonate roofing panels
30 × 24" clear acrylic panel
Roofing screws with EPDM washers

(2) 3½" exterior-grade butt hinges with screws
(2) ¼ × 4" eyebolts
3½ × 5/16" stainless-steel machine bolts
 (2 bolts with 8 washers and 2 nuts)
(2) Heavy-duty hook-and-eye latches
Outdoor thermometer with remote sensor (optional)
Eye and ear protection
Work gloves

How to Build a Jumbo Cold Frame

Drill pilot holes and fasten the frame end pieces between the frame side pieces with 3" deck screws to create the rectangular frame. Do the same with the base pieces to create the base. Use two screws for each joint.

Stabilize the corners of the canopy frame with braces cut to 45° angles at both ends. Install the braces on-the-flat, so their top faces are flush with the tops of the canopy frame. Drill pilot holes and fasten through the braces and into the frame with one 2½" screw at each end. Then, drive one more screw through the outside of the frame and into each end of the brace. Check the frame for square as you work.

Assemble the canopy glazing framework using ½" PVC pipe. Cut all the ribs 37" long. You can cut these easily with a miter saw, hacksaw, or jigsaw.

Use 2" deck screws as receptors for the PVC pipes. Drive the screws in 1" from edge and ¾" from the ends, angling the screws at about 35 to 45° toward the center. Leave about ¾" of the screw exposed. Drive two additional screws in at 32¼" from each end.

Install the PVC ribs by putting one end over the 2" screw, then curving the PVC until the other end fits over the opposite screw. Take your time with this, and use a helper if you need. **Note:** Hopefully you've remembered to buy the flexible PVC, not the Schedule 40 type used for indoor plumbing.

(continued)

How to Build a Jumbo Cold Frame (continued)

Hold up and mark a smooth piece of clear acrylic for the end panels. The clear acrylic should cover the 2 × 3 and follow the curving top of the PVC. Cut the clear acrylic with a plastic-cutting jigsaw blade.

Drill ¼" holes along the bottom of both panels about ⅝" up from the edge of the panel. Space the holes 2½" from ends, then every 16". Also mark and drill rib locations on the roof panels about 6" up from bottom, spacing the holes at 1⅝" and 33¼" from each end. Install the panels 1½" up from the bottom of the 2 × 3 with the roofing screws. The ends of the panels should extend 1" beyond the 2 × 3s.

Adjust the PVC ribs until the predrilled holes in the roof panels are centered on them, then predrill the PVC with a ⅛" bit. Fasten the panels to the two center ribs.

Lap the second sheet over the first, leaving roughly the same amount of panel hanging over the 2 × 3. Fasten the second sheet the same way as the first. Insert filler strips at each end under the polycarbonate, then drill through those into the PVC ribs. Now add additional screws about every ⅛". You can just predrill the holes with the ⅛" bit (the polycarbonate panels are soft enough that the screws will drive through them without cracking).

Set the clear acrylic end panels in place, butting them against the filler at the top. Mark screw locations. Place the panel on a piece of plywood and predrill with a ¼" diameter bit to avoid cracking the clear acrylic, which isn't as soft or flexible as the polycarbonate. Screw the panels in place with roofing screws, hand-tightening with a screwdriver to avoid cracking the clear acrylic. Don't overtighten.

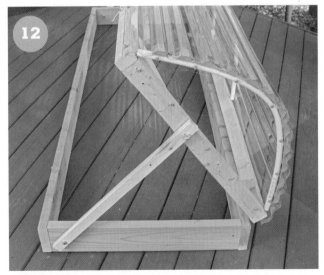

Mount the canopy to the cold frame base with two exterior hinges. The canopy frame should fit flush over the base on all sides. Screw in two hook-and-eye latches in front.

Attach a prop stick to each side with a stainless-steel bolt and nut. Insert three washers (or more) between the prop stick and the 2 × 6 base so the prop stick clears the clear acrylic side panel. Drill a few additional 5/16" holes in the stick and the frame for the eyebolts, so that you can prop the canopy open at different heights. Now, prepare the ground and place the cold frame in the desired location. Anchor the base to the ground using 16" treated stakes or heavy-duty metal anges driven into the ground and secured to the frame.

planting trees

26

Wind saps heat from homes, forces snow into burdensome drifts, and can damage more tender plants in a landscape. To protect your outdoor living space, build an aesthetically pleasing wall—a "green" wall of tress and shrubs— that will cut the wind and keep those energy bills down. Windbreaks are commonly used in rural areas where sweeping acres of land are a runway for wind gusts. But even those on small, suburban lots will benefit from strategically placing plants to block the wind.

Windbreaks

Trees or shrubs planted in a row are aesthetically pleasing, plus they create a windbreak that muffles noise, provides privacy, and deflects harsh winds and drifting snow from your home and yard. Plant the windbreak at a right angle to the prevailing winds.

Essentially, windbreaks are plantings or screens that slow, direct, and block wind from protected areas. Natural windbreaks are comprised of shrubs, conifers, and deciduous trees. The keys to a successful windbreak are: height, width, density, and orientation. Height and width come with age. Density depends on the number of rows, type of foliage, and gaps. Ideally, a windbreak should be 60 to 80 percent dense. (No windbreak is 100 percent dense.) Orientation involves placing rows of plants at right angles to the wind. A rule of thumb is to plant a windbreak that is 10 times longer than its greatest height. And keep in mind that wind changes direction, so you may need a multiple-leg windbreak.

A stand of fast-growing trees, like these aspens, will create an effective windbreak for your property just a few years after saplings are planted.

○ Creating a Windbreak

TOOLS & MATERIALS

Shovel

Garden hose

Utility knife

Trees

Soil amendments (as needed)

Eye and ear protection

Work gloves

Wind saps heat from homes, forces snow into burdensome drifts, and can damage more tender plants in a landscape. To protect your outdoor living space, build an aesthetically pleasing wall—a "green" wall of trees and shrubs—that will cut the wind and keep those energy bills down. Windbreaks are commonly used in rural areas where sweeping acres of land are a runway for wind gusts. But even those on small, suburban lots will benefit from strategically placing plants to block the wind.

Essentially, windbreaks are plantings or screens that slow, direct, and block wind from protected areas. Natural windbreaks are comprised of shrubs, conifers, and deciduous trees. The keys to a successful windbreak are height, width, density, and orientation. Height and width come with age. Density depends on the number of rows, type of foliage, and gaps. Ideally, a windbreak should be 60 to 80 percent dense. (No windbreak is 100 percent dense.) Orientation involves placing rows of plants at right angles to the wind. A rule of thumb is to plant a windbreak that is ten times longer than its greatest height. And keep in mind that wind changes direction, so you may need a multiple-leg windbreak.

Evergreens and deciduous trees both are effective windbreaks. These balsam trees grow only about 1' per year, but some faster-growing species exceed 10' per year in new growth.

Windbreak Benefits

Windbreaks deliver many benefits to your property.

Energy conservation: reduce energy costs by 20 to 40 percent.

Snow control: single rows of shrubs function as snow fences.

Privacy: block a roadside view and protect animals from exposure to passers-by.

Noise control: muffle the sound of traffic if your pasture or home is near a road.

Aesthetic appeal: improve your landscape and increase the value of your property.

Erosion control: prevent dust from blowing; roots work against erosion.

How to Plant a Windbreak

Before you pick up a shovel, draw a plan of your windbreak, taking into consideration the direction of the wind and location of nearby structures. Windbreaks can be straight lines of trees or curved formations. They may be several rows thick, or just a single row. If you only have room for one row, choose lush evergreens for the best density. Make a plan.

Once you decide on the best alignment of trees and shrubs, stake out reference lines for the rows. For a three-row windbreak, the inside row should be at least 75' from buildings or structures, with the outside row 100 to 150' away. Within this 25 to 75' area, plant rows 16 to 20' apart for shrubs and conifers and no closer than 14' for deciduous trees. Within rows, space trees so their foliage can mature and eventually improve the density.

Dig holes for tree root balls to the recommended depth (see pages 86 to 87). Your plan should arrange short trees or shrubs upwind and taller trees downwind. If your windbreak borders your home, choose attractive plants for the inside row and buffer them with evergreens or dense shrubs in the second row. If you only have room for two rows of plants, be sure to stagger the specimens so there are no gaps.

Plant the trees in the formation created in your plan. Here, a row of dwarf fruit trees is being planted in front of a row of denser, taller evergreens (Techny Arborvitae).

food preparation & preservation

We depend on our food, just as we do our air and water, to be healthy and wholesome. But increasingly, we have no idea where our food comes from and what is done to it before it reaches us. The farther we move from the source of our food, the more it needs to be processed to get to us. Although shelf life is increased, much can be lost.

That's why there's something wholly satisfying about raising and storing your own food. It's not just that it's amazingly healthy—that much should be obvious. More than that, it's that you're taking steps to ensure that there are no toxins in your family's food. You're also taking empty calories out of your diet—like when you replace soda with fresh-pressed cider. But beyond health, there's the benefit of more flavorful edibles.

Grow heirloom tomatoes and you've given yourself the gift of rich, beautiful, incredibly tasty vegetables that can't be rivaled by what's on offer at the local grocery. Can and preserve part of that harvest, and you'll be enjoying homemade marinara and stews through the winter, while other people are buying factory-canned soups full of preservatives and who knows what else.

But there's another side to self-sufficient food growing, harvesting, and preparation. Once you get outside and start pressing cider, drying fruit, or making your own cheese, you'll discover the vastly enjoyable pleasure of crafting your own wholesome food. The act of walking down a grocery store aisle simply can't compete with slicing up your own fresh-picked cabbage to make a family-favorite cole slaw, or hanging herbs just cut from your garden to make dried additions to your next soup. Get in the habit of tending your plants and working with the harvest they produce, and it quickly becomes the farthest thing from work.

preserving your bounty

Preserving garden produce does not need to be an overwhelming task, but it is important to think through your strategy before you get started. To understand how to preserve food, it's important first to understand why fresh fruits and vegetables spoil and decay. There are two main culprits: First, external agents, such as bacteria and mold, break down and consume fresh food. Second, naturally occurring enzymes—the very same ones that cause fruits and veggies to ripen—are also responsible for their decay. Canning, freezing, drying, and cold storage (along with salt-curing/smoking, and making fruit preserves) are all ways to slow down or halt these processes while retaining (to varying degrees) nutrition and taste.

Each method for preserving food has its strengths and weaknesses. It's important to weigh these carefully and decide which preservation method is the best fit for your garden's needs and your lifestyle.

Canning is one of the most popular forms of food preservation. The process is reliable and can be applied to a wide variety of vegetables.

TIP

Preserving Methods

Canning. Canning changes the taste of foods and results in some vitamin loss, but is versatile and can be used to preserve many different kinds of foods for lengthy periods of time.

Cellaring. Live storage preserves produce with minimal effect on taste or nutritional value, but it only makes sense for a few foodstuffs. The fruits and vegetables that can be cellared for limited periods of time require a storage environment that must be carefully regulated.

Freezing and drying. Freezing and drying both retain a high percentage of vitamins and have a minimum effect on flavor, but only certain foods can be preserved with these methods and careful preparation and regulation is extremely important. Freezing can damage the cellular structure of fruits and vegetables, causing an unpleasant mushiness that makes them suitable only for dices and purees. Drying also changes the nature of fruits and vegetables; for example, drying fruits causes the caloric value to double or triple as the starches convert to sugars during the process.

Choosing the Best Preservation Method

Produce	Canning	Freezing	Dehydration	Live Storage
Apples	✔		✔	✔
Asparagus	✔	✔		
Beans (green)	✔	✔		
Beans (lima)	✔	✔	✔	
Beets	✔			✔
Broccoli		✔	✔	
Brussel Sprouts			✔	✔
Cabbage				✔
Carrots	✔	✔		✔
Cauliflower		✔		
Celery				✔
Cherries	✔		✔	
Corn	✔	✔		
Cucumbers	✔ (with pickling)			
Onions				✔
Pears	✔		✔	✔
Peaches	✔	✔	✔	
Peas	✔	✔	✔	
Peppers (green)		✔	✔	
Peppers (hot)			✔	
Potatoes	✔			✔
Pumpkin	✔			✔
Radishes				✔
Spinach	✔	✔		
Squash (summer)	✔	✔	✔	
Squash (winter)	✔			✔
Strawberries		✔	✔	
Tomatoes	✔	✔	✔	

○ Preservation by Freezing

Freezing is the best way to preserve delicate vegetables. It's also a quick process that is perfectly suited to smaller batches of food. In this process, foods should be blanched to stabilize nutrients and texture, cooled to preserve color, packaged in an airtight container, and frozen as quickly as possible. Frozen food, if properly packaged and contained within a temperature-consistent frozen environment, can be preserved for as long as a year. Of course, the longer you wait to eat your food, the more it will break down, which results in a loss of taste and freshness. Food can also absorb ambient flavors in the freezer environment, negatively affecting the taste.

Generally, the colder you keep your freezer, the longer your frozen food will stay tasting fresh. For best results, use a chest freezer instead of the little box above your refrigerator. Although chest freezers are an investment, they maintain colder temperatures more consistently than your refrigerator's freezer. The ideal temperature for your chest freezer is −5 degrees Fahrenheit, and it should be no warmer than 0 degrees. Even a few degrees above zero will cut the freezer life of your food in half.

○ Preservation by Drying

There are many advantages to dehydrating produce from your garden. Most dehydration methods require very little extra energy other than that already provided by the sun. Also, dehydrated foods, if prepared correctly, retain much of their original beauty and nutritional value. And since foods lose so much of their mass during the dehydration process, they do not require much space to store through the winter and can easily be rehydrated to taste delicious months after the harvest.

Apples are a favorite fruit for drying because they retain so much of their flavor. Look for sweet varieties like Fuji. Core them and cut them into ⅛"-thick rings or slices for drying. Peeling is optional. Dip the apples in lemon juice immediately after cutting or peeling to prevent browning.

Dehydration is a food preservation technique that has been used for centuries all around the world. Removing 80 to 90 percent of the moisture in food, it halts the growth of spoilage bacteria and makes long-term storage possible. Warm, dry air moving over the exposed surface of the food pieces will absorb moisture from the food and carry it away. The higher the temperature of the air, the more moisture it will absorb, and the greater the air movement, the faster the moisture will be carried away.

Temperature matters a lot in food drying—air at a temperature of 82 degrees will carry away twice as much moisture as air at 62 degrees. This process also concentrates natural sugars in the foods. The faster the food is dried, the higher its vitamin content will be and the less its chance of contamination by mold. Extremely high temperatures, however, will cause the outside surface or skin of the food to shrivel too quickly, trapping moisture that may cause spoilage from the inside out. Exposure to sunlight also speeds up the drying process but can destroy some vitamins in foods.

Often, foods should be treated before drying. Blanching as you would for freezing is recommended for just about any vegetable (notable exceptions being onions and mushrooms). Some fruit and vegetables dry best if cut into pieces, whereas others should be left whole. Coating the produce can help preserve the bright color of skins. Many dipping mixtures may be used (consult a recipe book), but lemon juice is probably the most common.

○ Preservation by Home Canning

Canning is a traditional method for preserving produce. It is not difficult to master, but it's important to pace yourself. Try not to plan more than one canning project a day to keep the work manageable and enjoyable. Also, make sure you are familiar with how to use your canning equipment safely, and that you have a reliable recipe to reference for each food you plan to can. Every fruit and vegetable has a different acidity and requires slightly different accommodations in the canning process.

To get started with canning, there are two main tools to become familiar with: a water bath canner and a pressure canner. Foods with high acidity, such as fruits (including tomatoes), can be canned in a boiling water bath. Less acidic foods, including most vegetables, and any combination of high- and low-acidity foods must be processed using a pressure canner. Water bath and pressure canners are NOT interchangeable, largely because they reach vastly different temperatures during their processes. Always make sure the canner you use is appropriate for the produce you're preserving and follow your canning recipe exactly.

Other tools you'll need include canning jars, measuring cups, a long-handled spoon, a funnel, a jar lifter, and cooking pots. Canning jars typically have two-piece metal lids: the metal band can be reused whereas the disc part of the lid cannot form an adequate seal more than once, and should be discarded after one use. Always inspect jars carefully before beginning. Check for nicks on the rim or cracks anywhere in the jar. Discard or repurpose any imperfect jars as they will not be able to form an adequate seal.

TIP

Beware of Botulism

Home canning is perfectly safe if all instructions are followed exactly. However, if canning procedures are not followed, harmful bacteria can fester while your produce sits in your pantry waiting to be used. The best known of these bacteria is *clostridium botulinum*, which produces a potent toxin that is odorless, colorless, and fatal to humans in small amounts. Cases of botulism poisoning are rare, but to avoid this toxic substance, it's important to always follow the home canning recipe and procedures exactly. And if in doubt, throw it out.

Make sure you understand how to use your home canning equipment before you get started. Take time to read through the manual that comes with your canner, and make sure you use the right type of home canner for the fruit or vegetable you're planning to preserve.

The Home Canning Process

1. **Wash and heat the jars.** Immerse jars in simmering water for at least 10 minutes or steam them for 15 minutes. Heat jar lids (just the disc part) in a small saucepan of water for at least 10 minutes. Keep lids hot, removing one at a time as needed.

2. **Pack food in the jars.** Different packing methods are used for different types of produce. In cold packing, raw food is placed in a hot jar and then hot liquid is poured over the food to fill the jar. In hot packing, foods are precooked and poured into a hot jar immediately after removing them from the heat source.

3. **Watch your headspace.** Headspace is the amount of space between the rim of the jar and the top of the food and is very important to making sure that your canning jars seal correctly. Always follow your recipe's directions—generally it's best to leave about 1 inch of headspace for low-acid foods, ½ inch for acidic foods, and ¼ inch for pickles, relishes, jellies, and juices.

4. **Remove air bubbles.** Insert a nonmetal spatula or chopstick and agitate the food to remove all air bubbles.

5. **Place the lid.** Clean the jar rim, then set a hot disc on the jar rim and screw on the band until you meet the initial point of resistance and no further.

6. **Heat.** Place jars on the rack in the water bath or pressure canner and process immediately. Follow the directions for your canner.

7. **Cool.** Allow the jars to cool slowly after processing—cooling too quickly can cause breakage. Typically, jars should cool along with the water they're submerged in, but follow the directions for your canner. Do not tighten the lids unless they are very loose. As the jars cool, you'll hear them "pop" when they are properly sealed. If the jar does not seal, refrigerate and eat within the next couple days.

8. **Clean and label.** After cooling and confirming the jar's seal, wash the outside of the jar and label with the content and date.

9. **Store.** Store in a cool, dark cupboard or pantry. If a jar loses its seal during storage (i.e., if the metal disc does not pop when you remove it), the food inside is not safe to eat. Dump it on the compost bin and try a different jar.

Canned Food Safety Quiz

Ask yourself the following 10 questions to determine if your home-canned food is safe to eat:

1. Is the food in the jar covered with liquid and fully packed?
2. Has proper headspace been maintained?
3. Is the food free from moving air bubbles?
4. Does the jar have a tight seal?
5. Is the jar free from seepage and oozing from under the lid?
6. Has the food maintained a uniform color?
7. Is the liquid clear (not cloudy) and free of sediment?
8. Did the jar open with a clear "pop" or "hiss" and without any liquid spurts?
9. After opening, was the food free of any unusual odors?
10. Is the food and underside of the lid free of any cottonlike growths?

If you can answer "yes" to all of the questions, your food is probably safe. That said—if you have even a small suspicion that a jar of food is spoiled—dump it in the compost bin. Never, under any circumstances, taste food from a jar you suspect may have spoiled or lost its seal. Botulism spores have no odor, cannot be seen by our eyes, and can be fatal, even in small doses.

◦ Drying Produce Indoors

Drying vegetables indoors allows you to carefully control the drying conditions and offers more protection from insects and changes in weather. An electric food dehydrator appliance is the simplest choice for indoor drying. If you don't have a dehydrator, the next best option is in or around your oven, although any hot, dry area will do—possibly even your attic or the area around a heater or cookstove.

If you plan to dry produce in your oven, keep in mind that the process typically takes 8 to 12 hours. Preheat your oven and check that it can maintain a temperature of 130 to 145 degrees for at least an hour—some ovens have a difficult time holding low temperatures like this, and going over 150 degrees can be disastrous for drying produce. Wash and prepare the food, then spread food in single layers on baking sheets, making sure the pieces do not touch. Place the sheets directly on the oven racks, leaving at least 4 inches above and below for air circulation. Also, make sure to leave the oven door slightly ajar to allow moisture to escape. Rearrange the trays and shift food from time to time to ensure even drying.

You may also dry food on your oven's range by creating a chafing dish. To create a chafing dish on your range, you'll need two baking trays: The first must be large enough to cover all burners and hold a 3-inch-deep reservoir of water. The second tray should fit on top of the first. Fill the bottom tray with water and set all burners to low heat. Throughout the process, refill the reservoir periodically to make sure food doesn't burn, and move/turn food as necessary to ensure even drying. Place a fan nearby to keep the air moving around the room, which will help carry moisture away from the food more quickly.

When drying produce in the oven, leave the door slightly ajar to allow moisture to escape, and carefully monitor temperature to ensure the oven doesn't heat to over 150°.

An electric produce dehydrator can dry large quantities of fruits or vegetables in a sanitary environment. The stackable trays allow you to match the appliance's drying capacity to your needs each time you use it.

How Long Does Dehydration Take?

Drying times vary considerably—from a few hours to many days, depending on the climate, humidity, drying method, and the moisture content of the food you're dehydrating. Generally, fruit is done drying if it appears leathery and tough and no moisture can be squeezed from it. Vegetables should be so brittle and crisp that they rattle on the tray. To check for completed dehydration, you can also check the food's weight before and after the process. If the food has lost half its weight, it is two-thirds dry, so you should continue to dry for half the time you've already dried.

To double-check that your food is dry, place it in a wide-rimmed, open-topped bowl covered with cheesecloth fastened with a rubber band. Place the bowl in a dry place, and keep the food in the bowl for about a week. Stir it a couple times a day—if any moisture or condensation appears, you should continue to dehydrate.

Pasteurization & Storage

Regardless of the drying method used, food should be pasteurized before storage to ensure that there are no insect eggs or spoilage microorganisms present. To pasteurize, preheat the oven to 175 degrees. Spread dried food 1 inch deep on trays and bake in the oven for 10 to 15 minutes. Dried food is best stored in clean glass jars or plastic bags in a cool, dry place. Never store dried food in metal containers and carefully monitor the humidity of the storage environment. Containers should have tight-fitting lids and should be stored in a dark, dry place with an air temperature below 60 degrees.

TIP

Enjoying Your Dehydrated Food

Many foods are delicious and ready to eat in their dried forms—especially tomatoes and berries. But dried food can also be rehydrated before eating. To rehydrate food, pour boiling water over it in a ratio of 1½ cups of water to 1 cup of dried food, then let the food soak until all the water has been absorbed. You may also steam fruit or vegetables until rehydrated. Rehydrated vegetables should be cooked before eating, whereas rehydrated fruits are acceptable to eat without cooking after rehydration.

After drying and before storing, use your oven to heat fruits and vegetables to a high enough temperature to kill bacteria and related contaminants. About 15 minutes at 175°F will suffice for most produce, provided it is not in layers over 1" deep. Oven-drying takes about a half day at 140° or so.

solar dryer

28

A solar dryer is a drying tool that makes it possible to air-dry produce even when conditions are less than ideal. This dryer is easy to make, lightweight, and is space efficient. The dryer makes a great addition to your self-sufficient home, allowing you to use your outdoor space for more than gardening. The dryer, which is made of cedar or pine, utilizes a salvaged window for a cover.

But you will have to adjust the dimensions given here for the size window that you find. The key to successful solar drying is to check the dryer frequently to make sure that it stays in the sun. If the air becomes cool and damp, the food will become a haven for bacteria. In a sunny area, your produce will dry in a couple days. Add a thermometer to the inside of your dryer box, and check on the temperature frequently—it should stay between 95 and 145 degrees Fahrenheit. You may choose to dry any number of different vegetables and fruits in the dryer, such as:

- Tomatoes
- Squash
- Peppers
- Bananas
- Apples

A Simple Dryer for a Hot Climate

If you live in an area with clean air and a sunny, hot, dry climate, you can simply load food onto a drying tray or rack and place it out in the sun on blocks so that air circulates around it. Cover the food with cheesecloth held a few inches off the food on sticks to keep insects away, and bring the tray indoor at nights. Drying will take two or three days.

An old glass window sash gets new life as the heat–trapping cover of this solar dryer.

Building a Solar Dryer

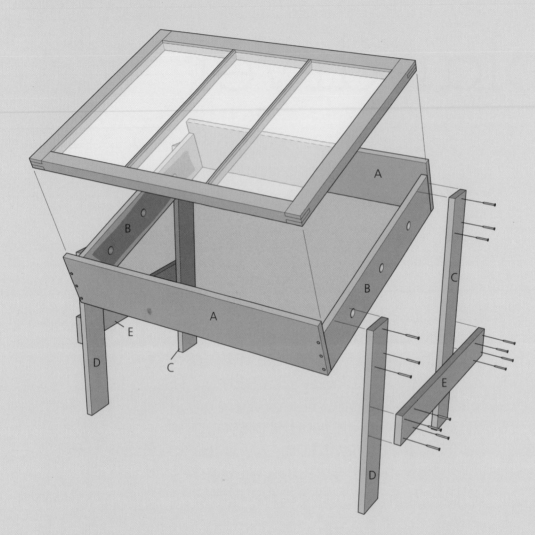

CUTTING LIST

Key	Part	No.	Dimension	Material
A	Front/back	2	¾ × 7½ × 34¾"	Cedar
B	Side	2	¾ × 5½ × 27⅛"	Cedar
C	Leg (tall)	2	¾ × 3½ × 30"	Cedar
D	Leg (short)	2	¾ × 3½ × 22"	Cedar
E	Brace	2	¾ × 3½ × 24"	Cedar

TOOLS & MATERIALS

1" spade bit

Circular saw

(1) 1 × 8" × 8'

(1) 1 × 6" × 8'

Eye protection

(2) 1 × 4" × 8'

Stapler

1¼" deck screws

Drill

Staples

Insect Mesh—fiberglass 28⅞ × 34¾"

Window sash

1½" galv. finish nails

Brad nails

Eye and ear protection

Work gloves

How to Build a Solar Dryer

Assemble the box. Attach the wider boards for the frame by driving screws through the faces of the 1 × 8" boards into the ends of the 1 × 6" boards. There will be a difference in height between these pairs of boards so that the window sash can sit flush in the recess created.

Install the mesh. Staple the screen to the frame. Then tack the retainer strips over the screen to the frame with 3-4 brad nails per side. Trim off the excess mesh.

Build the stand. Attach each 24" board to a 30" board (in the back) and a 22" board (in the front) with 1¼" deck screws. Then attach the finished posts to the frame with three 1¼" deck screws in each post.

Drill three 1" holes for ventilation in each 1 × 6" board equally spaced along the length of the board, leaving 5" of room on each end for the posts. Staple leftover insect mesh behind the ventilation holes on the inside of the frame.

Finish the project by sliding the window sash into place.

solar fruit dryer

This all-metal drying rack has more of an industrial look than most homespun wood versions, but there are good reasons for opting for metal. First, when you're working with food, you want a material that is washable and heat-resistant. Metal meets both of these requirements, while wood really meets neither, at least not over the long haul. Also, when you're dealing with extended exposure to sunlight, you want a material that's UV-resistant and won't warp, dry out, or deteriorate. Metal wins again.

Buying new materials for this metal dryer may cost more than using old scrap wood and some window screening, but the slightly higher price is the only drawback. Building the dryer is easy and straightforward (saving you time), and the dryer will outlast any appliance in your house (saving money in the long run).

You can customize the size of the dryer to fit your needs and drying locations. The "shelves" are stainless-steel cooling racks, the kind used by home bakers for cooling cakes and cookies. Buy as many racks as you like in the desired size, then build the dryer framework to fit the racks. Choose grid-style racks with thin wires and the smallest openings available (½-inch or smaller). The metal reflector at the bottom of the dryer reflects heat up toward the racks to speed the drying process.

When determining the size of your dryer, consider the amount of food you're likely to dry at one time, as well as the interior capacity of your oven. Solar dryers are designed to use the sun for power, of course, but it never hurts to have the option of slipping the dryer into the oven—for a little extra drying or just as a handy place to store the dryer at night.

Drying is a perfect way to preserve a large fruit harvest. It's also a very easy process when you have a fruit dryer that taps the power of the sun, like the one in this project.

TIP

Tips for Solar Drying

Although it seems like a very simple process, solar drying fruit is not just a case of drying out something that is juicy. If not done correctly, you can wind up with a big harvest of inedible leathery pieces.

• Start with fresh, fully ripened—but not overripe—fruit or vegetables.
• Cut foods evenly for consistent drying; some foods such as figs, grapes, and plums should be pierced, not sliced, while others can be sliced or chopped for drying.
• Pre-treating fruits in a solution of ascorbic acid, citric acid, or lemon juice can minimize discoloration and help prevent unwanted bacterial growth.
• Rotate racks as needed so foods dry at roughly the same time.
• If insects are a problem, cover the dryer loosely with a piece of flexible window screen, cheesecloth, or even mosquito netting; just be aware that this will block some sunlight and airflow, adding to the drying time.
• Pasteurize dried foods before storing by briefly heating them to 175°F (see page 181 for details).
• Store dried food in airtight glass or plastic containers kept in a cool, dark place.
• Contact a local extension office for tips and recommendations on drying foods specific to your region.

Building a Solar Dryer

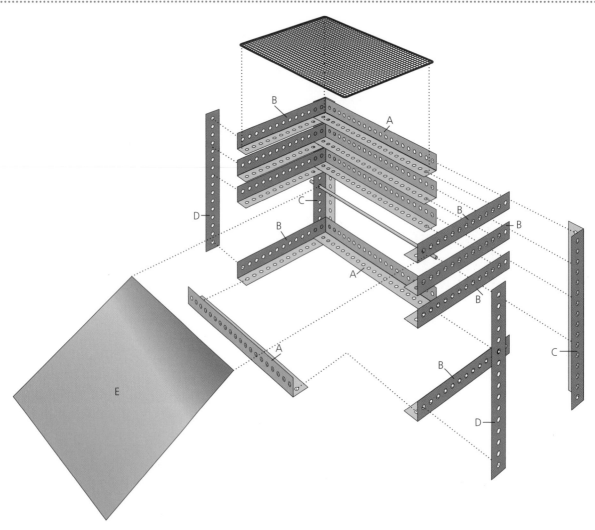

CUTTING LIST

Key	No.	Part	Dimension	Material
A	5	Shelf (front/back)	16½"	1½ × 1½" slotted steel angle
B	8	Shelf (side)	10½"	1½ × 1½" slotted steel angle
C	2	Rear upright	13⅛"	1½ × 1½" slotted steel angle
D	2	Front upright	13⅛"	1½ slotted steel flat bar
E	1	Reflector	16½ × 12"	Sheet Metal

TOOLS & MATERIALS

Marking pen
Reciprocating saw or jigsaw with metal-cutting blade, or hacksaw
Metal file
Wrenches
Tin snips
Stainless-steel cooling racks
(5) 4' × 1½ × 1½" or 1¼ × 1¼" slotted steel angle*
(1) 3' × 1½" or 1¼" slotted steel flat bar*

(20) ¾ × ⁵⁄₁₆" hex bolts with washers and nuts (sized for holes in steel angle and flat bar)
Hardwood dowel (sized for holes in steel angle)
Scrap wood and screws (as needed)
Eye and ear protection
Work gloves

*Use plain-steel angle and bar, not galvanized or zinc-plated.

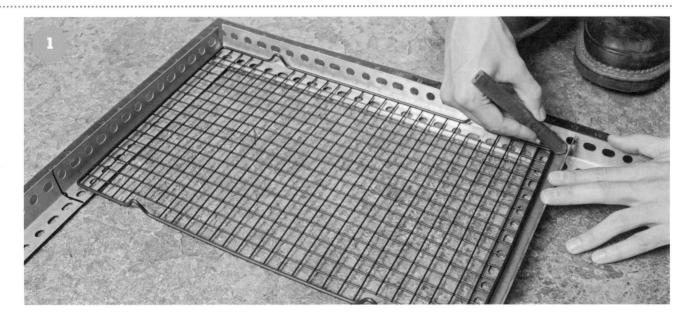

Design the dryer frame to fit your cooling racks. Add ½" to ¾" to the width and depth of each shelf space to make sure the racks have enough room to slide easily along the shelf angles. Also allow space for the heads of the bolts that hold the frame together. Lay out the cuts for the two rear uprights on the steel angle; mark the cuts for the two front uprights on the flat steel bar. The lengths of these pieces are all the same and should equal the overall height of the dryer minus the thickness of the angle material (the front and lower rear crossbars will sit below the uprights). **Note:** The trick to laying out the cuts is to align the holes in the corresponding pieces so that all of the holes are lined up in the finished assembly. You can mark and cut each piece as you go, using the cut piece as a template for aligning the holes and marking the cut for the next piece.

File the ends of each piece to remove any burrs and sharp edges. Mark and cut two shelf angles and the upper rear crossbar, again making sure the holes are aligned between like pieces.

Cut the four uprights. You can cut the pieces with a hacksaw, reciprocating saw, or jigsaw with a metal-cutting blade, although a cut-off saw (a metal-cutting chop saw, not a standard miter saw) is ideal.

(continued)   189

Building a Solar Dryer (continued)

Drying Your Harvest . . . and Then Some

Once you've completed your new solar fruit dryer, the hard part's over. That's because the process of drying fruits and vegetable could not be simpler. Just cut and arrange whatever it is you need to dry, and let the sun do the work. Keep the dryer in direct sun as much as possible for best results. Most fruits will dry completely in one day. Other sturdier foods, such as tomatoes, should be dried for two days.

Don't limit yourself. The list of potential fruits and vegetables for drying is a long one, and may even spur you to plant and grow new vegetables and fruits. A short list of some of the more unusual dried fruits and veggies includes:

- Heirloom tomatoes
- Plantains
- Yellow onions
- Mushrooms (wild and otherwise)
- Zucchini (and just about any other squash)

You can even use the dryer to dry herbs in a pinch (although you'll find it easier to use the Herb Dryer on page 193). The idea is the same either way: have dried food ready to use out of season. All you need to do to bring the dried vegetables back to life is water and time. Simply soak them for 10 to 30 minutes in warm water or oil.

Of course, you can also eat anything you dried in its dried state. Most dried fruits add an incredible element sprinkled over cereal or on top of a bowl of yogurt. Eat sundried tomatoes plain, with sprinkling of salt and a drizzle of avocado oil.

Temporarily assemble the uprights, base, and one or two crossbars to make sure that everything fits well, the holes are properly aligned, the fruit dryer is square, and the rack fits. The bolt heads protrude less than the nuts, so they should go on the inside (or shelf-side) of the frame; washers and nuts go on outside of the uprights and the underside of the crossbar.

If everything goes together properly, cut the remaining pieces, making sure to line up the holes. The easiest way to do this is to use one piece as a template for length and hole locations. If you have a good saw and sharp metal cutting blade, you can cut several pieces at once. Use a file to smooth the metal edges.

6

Complete the dryer-frame assembly, hand-tightening the bolt connections. Then tighten each bolt connection with wrenches. You may need to flip the cooling racks over if the legs get in the way.

7

Cut the sheet metal for the reflector panel to size with tin snips, or a jigsaw with a metal cutting blade if the metal is too thick for snips (sandwich the metal between thin pieces of plywood for a smoother cut). Insert a dowel through the rear uprights at the height desired, then rest one end of the panel on top of the dowel to reflect sunlight and radiate heat toward the racks.

Tip

You can change the angle of the reflector panel to catch the most sunlight or to follow the path of the sun during the day. If the metal is thin and too flexible, fasten it to a piece of plywood, securing the metal with small screws driven through the panel's top side or with silicone adhesive. You can also use plywood or even cardboard wrapped with foil.

herb-drying rack

30

Herbs are some of the easiest and most enjoyable plants to grow. They can be grown in a small garden plot, in a container garden, or even on a windowsill. The fact is, you can easily grow more than you need with very little effort. Of course, the self-sufficient gardener never lets something go to waste in the garden—and that means making the most of any extra herbs by drying them.

The drying rack described in the pages that follow will give you a place to dry a large harvest of several different herbs. In fact, it's really two drying racks in one. Herbs are dried in two ways: laying down or hanging up. Hardier herbs should be tied in bundles and hung to dry. These include lavender, rosemary, sage, thyme, and parsley and any herb with tough, woody stems. More delicate leaf herbs like basil, mint, and tarragon should be dried flat on a screen. To cover all the bases, the rack in this project includes three lines on top for hanging and three tray slots to hold drying screens. This should provide all of the space you'll need for even a large herb harvest.

We've also included instructions for making your own drying screens. These are basically old-fashioned wood window screens. They are fairly easy to make, and by crafting your own, you control the dimensions. However, you may prefer the ease of buying premade screens, which are available at home centers and hardware stores. If that's the case, simply alter the length of the braces and top to suit the screens you buy (as well as the height of the dadoes). Also, if you're purchasing screens, it's wise to invest in more durable aluminum, rather than fiberglass mesh. Choose aluminum frames for a wide selection of sizes or pick from a much smaller selection of wood-framed screens.

Combining hanging and screen drying, this rack can quickly and efficiently dry any herb you can grow. And it looks just as good as it works.

Building an Herb-Drying Rack

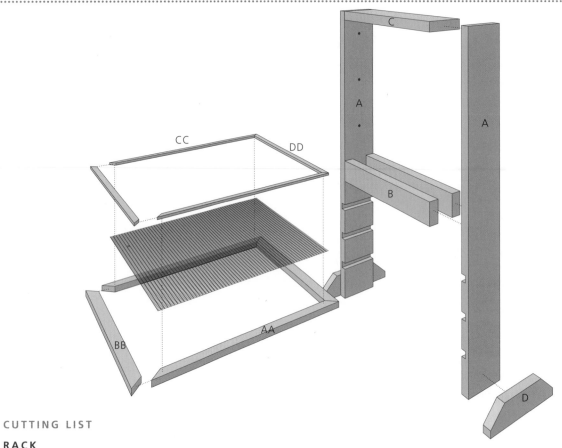

CUTTING LIST

RACK

Key	No.	Part	Dimension	Material
A	2	Sides	1½ × 5½ × 49"	2 × 6
B	2	Braces	1½ × 3½ × 26"	2 × 4
C	1	Top end side	1½ × 5½ × 26"	2 × 6
D	2	Feet end	1½ × 3½ × 12½"	2 × 4 pine

DRYING SCREEN

Key	No.	Part	Dimension	Material
AA	6	Frame end piece	¾ × 1½ × 26⅞"	1 × 2
BB	6	Frame side piece	¾ × 1½ × 20"	1 × 2
CC	6	Molding end	¼ × ¾ × 25⅜"	Wood Screen Molding
DD	6	Molding side	¼ × ¾ × 18½"	Wood Screen Molding

TOOLS & MATERIALS

Window screen mesh (aluminum or fiberglass)
16-gauge stainless-steel picture hanging wire
(6) 1³⁄₁₆" steel screw eyes
Miter saw
2½" screws

Router with ¾" straight bit
   (or table saw with dado blade)
#5 × ⅝" corrugated fasteners
Cordless drill and bits
Staple gun and staples

¾" brads
Clamps
Linseed oil
Eye and ear protection
Work gloves

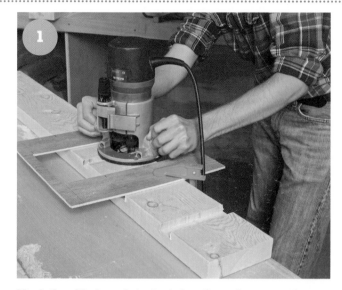

Mark the side boards for the dadoes. Set up the router with a ¾" straight bit and clamp a cutting guide to the workpiece. Start the first dado cut 6" up from the bottom, ½" deep. Move the cutting guide and make a second pass so the total width is ⅞", or about ⅛" wider than the wood used for the drying rack. (If you use aluminum screen frames, just add ⅛" to the thickness of the frames.) Cut the remaining dadoes. Repeat for the other side board. You can also make the cuts on a table saw using a dado blade.

Measure and mark the placement of the screw eyes on the inside faces of the side boards (same sides as the dadoes). The first pair should be positioned in the center of each side board, 4" from the top. Each of the other two will be positioned 8" below the screw eye above it. Predrill holes and then screw in the screw eyes until they are snug.

Cut and attach the feet with a single screw. Drill pilot holes for the braces, working on one side at a time. Screw the brace to one side panel, and then drill the pilot holes and screw the brace to the opposite side panel. Flip the frame and repeat for the second brace.

(continued)

Building an Herb-Drying Rack (continued)

Tie the picture wire to the screw eyes using pliers to twist the ends and secure the wire. Make sure the wire is tight enough to support the herbs without sagging.

TIP

Herb Drying

Drying herbs is a simple process but one that needs to be done correctly to ensure usable herbs. Start by picking the herbs you want to dry before the first buds on the herb begin to bloom. Once the plant flowers, the herb will usually become bitter. Pick only healthy whole leaves and stems; avoid wilted, yellowed or diseased parts of the plant.

- When drying herbs, keep the dryer out of direct sun, because direct sun can diminish the flavor of the herbs.
- Speaking of flavor, keep in mind when it comes time to use the herbs that dried herbs usually have much more intense flavor than the same herb fresh—often by a factor of four.
- Crumble your dried herbs and store them in airtight containers, in a cool, dry, and dark location.

Stand the dryer up on a level surface. Check for plumb and mark the proper position of each foot. Drill pilot holes and secure each foot firmly along the marked lines, with two to three additional screws.

Project Detail: How to Make the Drying Screen

Cut 45° miters in the ends of the frame and molding pieces. Check the fit and adjust as necessary.

Assemble the frame, fastening the pieces together with corrugated staples (inset) driven at each corner.

Fasten the screen with ⁵⁄₁₆" staples every few inches. Pull the screen tight as you staple. Cut away the excess screen when you're done stapling.

Position the molding on the inside edge of the frame. Nail it to the frame with brads. Repeat the process to construct the two remaining screens. Coat the frames with linseed oil or other non-toxic sealant.

root vegetable rack

31

Before the widespread use of refrigeration, one of the methods home gardeners used for preserving root vegetables was to spread them out on drying racks in cool, dry basements and root cellars. That basic method still works just fine, and it's a better than just dumping your bumper crop into bags or boxes, where they're more susceptible to mold and rot.

This classic drying rack can hold hundreds of potatoes, carrots, beets, and other garden vegetables and fruits, and it's dirt simple to build. Made from inexpensive common pine, you can leave it unfinished or wipe on a food-safe butcher block oil (make sure it says "Food-safe" or "Food-grade" on the label). You can easily make the rack larger or smaller than our version by changing a few lengths, or even make a smaller countertop version for your kitchen; just be sure to leave the gaps between the slats for air movement and to follow the design for the drawers. The drawers can be opened from either direction for easy access to the vegetables.

Because the drawers are made from individual slats instead of plywood, there are lots of pieces to cut, but most of them are the same length, and if you take a moment to set up a stop block and cut several pieces at a time, you can cut them all in minutes. A compressor and nail gun will speed up construction considerably, but you can also just use screws or hand-nail everything, as carpenters would have done back in the nineteenth century.

Keep your vegetables and apples dry and fresh with this classic pine drying rack.

Building a Root Vegetable Rack

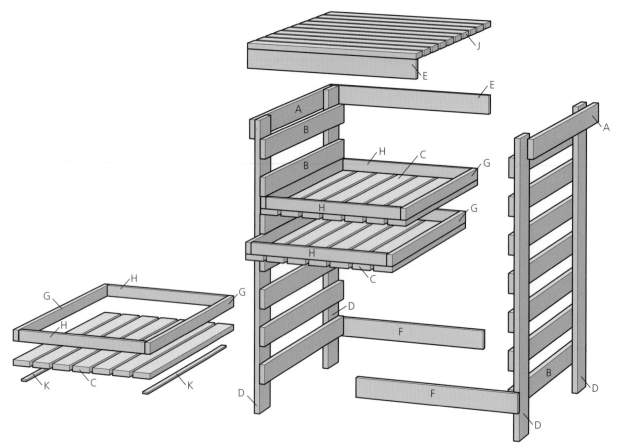

CUTTING LIST

Key	No.	Part	Dimension	Material
A	2	Top side trim	¾ × 2½ × 30 "	Pine
B	14	Drawer support	¾ × 2½ × 30 "	Pine
C	42	Drawer slat	¾ × 2½ × 30 "	Pine
D	4	Leg	¾ × 3½ × 38 "	Pine
E	2	Top front and back	¾ × 2½ × 23¾ "	Pine
F	2	Bottom stretcher	¾ × 2½ × 22¼ "	Pine
G	12	Drawer side	¾ × 1½ × 30 "	Pine
H	12	Drawer front and back	¾ × 1½ × 19 "	Pine
J	10	Top slat	¾ × 2½ × 23¾ "	Pine
K	12	Drawer guide	¼ × ¾ × 28 "	Pine screen mold

TOOLS & MATERIALS

Miter saw
Drill
Compressor and finish nailer
  or narrow crown stapler
Framing square
Clamps

Countersink bit
Wood glue
Self-tapping or drywall screws—1¼" and 2"
1½" finish nails (or narrow crown staples)
¾" brad nails
(21) 1 × 3 × 8' pine

(2) 1 × 4 × 8' pine
(7) 1 × 2 × 8' pine
(4) ¼ × ¾ × 8' pine screen mold
Sandpaper–150 grit
Eye and ear protection
Work gloves

Cut 56 lengths of 1 × 3 and 12 lengths of 1 × 2 to 30" (Parts A, B, C, G). To ensure that all the pieces are the same, set up a stop block wide enough so that you can cut several pieces at once. One simple way to make a stop block is to screw a long, temporary 1 × 3 fence to the miter saw fence, mark 30 from the blade, then screw on a stop block. Move the stop block as needed to cut the rest of the pieces.

Cut the legs and glue and nail on the bottom 1 × 3 drawer support. Keep the 1 × 3 square to the legs as you work by clamping a square in place. Glue each joint, nail it in place (if you have a pneumatic nailer), and then reinforce each joint with two 1¼" screws. Predrill with a countersink bit to avoid splitting the wood if you are hand-nailing.

Using a straight piece of 1 × 3 as a spacer, add the rest of the 1 × 3 drawer supports. Check the distance to the top frequently to make sure you're still square. Glue and nail each joint. Repeat these steps to build the second side.

(continued)

Building a Root Vegetable Rack (continued)

Stand the two sides up, top ends down, and join them with the top front and back pieces (E). Place the bottom stretchers (F) ¼ below the level of the first pair of drawer supports so that the drawer guides (K) on the bottom of the drawers will clear. Fasten all these joints with glue and predrilled screws for extra strength. Stand the drying rack right side up and set it aside.

Nail the 1 × 2 drawer sides together using a dab of glue to strengthen the joints. Make sure that the pieces are square.

Glue and nail the 1 × 3 bottom slats. Nail on the edge pieces first to keep the drawer square, then use ½ spacers (full ½ , not ½ plywood, which is slightly less than ½) to space the rest of the slats.

Sand the drawer bottoms with 150-grit sandpaper so they slide smoothly in the runners. Also sand the edges and corners to give the piece a more finished look.

Nail drawer guides to the bottom of each drawer so that they will stay on the tracks. Space the guides ¾" from each edge, using a piece of 1 × 3 as a guide. Remember to use ¾" brads to nail the guides. The guides are 28" so that they can be set back from the edges an inch, which makes them almost invisible.

The top is optional, but it gives the drying rack a more finished look, plus it creates more storage space. Cut the pieces using the stop block, then glue and nail them in place. Space the slats about $1^{1/16}$ apart, but you may need to adjust the spacing for the last few pieces.

root cellar

Cool storage areas such as cellars can be outfitted to keep produce fresh for the table throughout the long winter months. A wide variety of foods will stay fresh and delicious if stored in the right conditions—a space that is damp and cold, but not freezing. Typically, 32 to 40 degrees Fahrenheit is ideal for a root cellar environment. This type of food storage is entirely dependent on thermal mass and the natural cooling of outdoor air during the winter, and this isn't vulnerable to power outages. Traditionally, root cellars are an underground space built under or near the home, insulated by the ground and vented so cold air can flow in and warm air out in the fall. In the winter, the vents are then closed and the cellar maintains a cold—but not freezing—temperature, thanks to the earth's insulation.

Of course a walk-in root cellar built like this is the most reliable solution, but you can still practice cold storage without an external walk-in root cellar. The best systems are adapted to each home and climate, and can be as simple as a deep hole in the yard that is carefully covered, to a homemade basement cold room, like the one described on page 207.

A root cellar doesn't actually need to be underground, although they often are. Any cool, dark area will do.

How to Store Produce in a Root Cellar

Vegetable	Will Store For...	Special Instructions
Apples	5 to 8 months	Apples give off a gas that causes root vegetables to sprout or spoil, so store them in separate spaces. Apples also like to be moist—and store well in a barrel lined with paper or sawdust.
Cabbage and Cauliflower	Cabbage—3 to 4 months Cauliflower—1 month Chinese Cabbage—2 months	Store only sound, solid heads. Place the heads in plastic bags that have a few holes to let excess moisture escape. Remove roots and outer leaves.
Carrots and Parsnips	4 to 6 months	Snip off the leaves just above the crown. Store in covered containers filled with moist sand or moss.
Celery	2 to 3 months	Harvest just before heavy frost. Leave the roots and soil attached and set in moist sand in a shallow container on the cellar floor. The sand should be only deep enough to cover the roots and must be kept slightly moist. Cover or store in the dark.
Hot Peppers	6 months	Air dry and store in a cool, well-ventilated room.
Pears	3 months	Pack in loose paper in crates or barrels.
Potatoes and Other Root Vegetables	Beets—3 months Potatoes—5 to 8 months Turnips—4 to 5 months	Cut off the tops about ½" above the crown. Store bedded in vented plastic bags or covered crates filled with damp sphagnum moss or sand. Keep out of light. Note that potatoes like to be stored a little warmer—around 40°.
Pumpkins and Squash	Pumpkins—3 months Squash—6 months	Leave a part of the stem on each pumpkin or squash. Pumpkins and squash thrive in a slightly warmer space—between 55 and 60°. Keep pumpkins and squash dry—a humid space will cause them to deteriorate.
Onions and Garlic	Onions—8 months Garlic—7 months	Pull when tops fall over and begin to dry. When tops are completely dry, cut them off 1" from the bulbs. Cure for another week or two before placing in storage. Onions and garlic are best kept dry and stored in mesh bags or crates.

○ How to Set Up a Basement Root Cellar

Modern basements are typically too warm for long-term winter storage, but you can create an indoor version of a root cellar by walling off and insulating a basement corner and adding vents to the outside to let you regulate the flow of cold outside air into the insulated room. Your goal is to create a small room that is well insulated and will remain near freezing throughout the winter months. Cellar rooms are typically quite humid, so be sure to choose insulation materials that will hold up well in a moist environment.

First, choose a location for your cellar that is as far as possible from your furnace, and near a basement window, if possible. The window is a great place to install a vent—simply remove the window glass, replace it with insulated plywood, and run the vent through a hole in the wood. (You could also run a vent through a basement wall—as you would for a clothes dryer.) Choose a northeast or northwest corner location if you can. The more masonry surface in your root cellar room, the better—masonry walls provide thermal insulation to help create the ideal temperature inside. If a northeast or northwest corner won't work for your basement, choose the corner with the highest outdoor soil height.

Store only mature, high-quality vegetables in a root cellar: Small, cut, bruised, or broken vegetables will not store well and should be eaten right away. Check on your stored foods frequently to see how they're doing—if the vegetables begin to grow, the cellar is too warm. If they freeze, the cellar is too cold. If the skin starts to look dry or shriveled, the space is too dry. Remove decaying vegetables immediately to prevent rot from spreading to the rest of your food.

To store carrots: Cut off greens and wrap them in small groups of newspaper. Bury paper packages in dry sand.

Building a Root Cellar

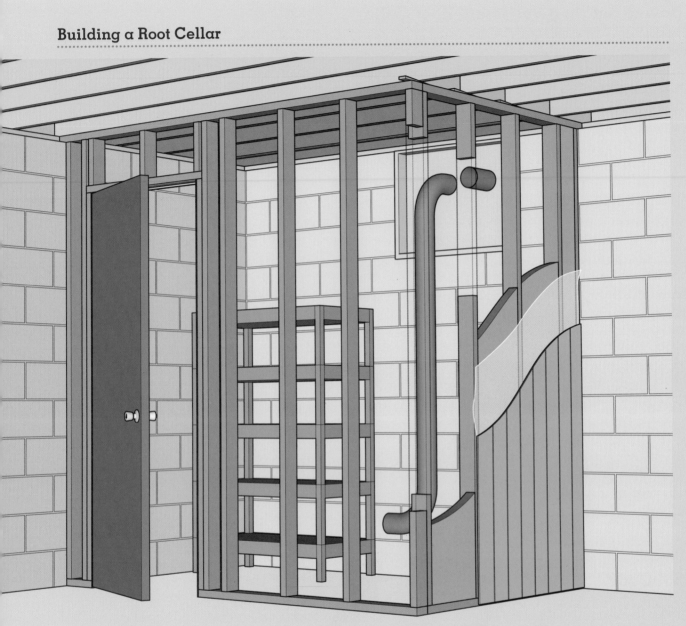

TOOLS & MATERIALS

Chalk	Drill	Stapler	Masking tape
Eye protection	Level	Stick-up light	4d finish nails
Ear protection	Deck screws	Fiberglass butt insulation	Glue
Construction adhesive	Circulating saw	Paneling or drywall	Wood screws
Concrete nails	Insect mesh	Steel garage service door	Framing square
Powder-actuated nail gun	Sheet—plastic vapor barrier	Weatherstripping	Sander
Louvered pipe vent cover	Foam insulation	Plywood window insert	4" vent duct or flexible
Caulk	Work gloves		dryer vent

Building a Root Cellar

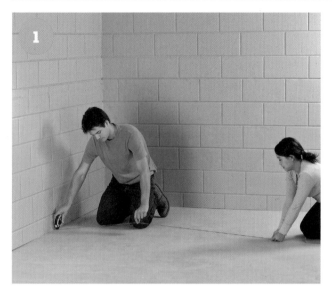

Outline the root cellar wall locations on your basement floor with chalk or a chalkline. Don't get too skimpy— the footprint should be at least 4 × 6' to make the project worthwhile.

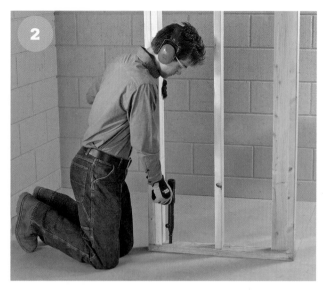

Build 2 × 4 stud walls with a framed rough opening for a door. Anchor the sole plates for the walls (use pressure-treated lumber) to the floor with construction adhesive and concrete nails driven into predrilled holes (or use a powder-actuated nail gun).

Blocking

Cap plate

Use a level to adjust the walls until they are plumb and then secure the cap plates of the framed walls to the joists above with deck screws. If the cap plate on the wall that's parallel to the joists does not align with a joist, you'll need to install wood blocking between the joists to have a nailing surface for the cap plate.

Insulate the interior walls to keep the ambient basement heat out of the root cellar. Rigid foam insulation is a great choice for root cellar walls, since it is more resistant to mold and deterioration from moisture than fiberglass batts.

(continued)

Building a Root Cellar (continued)

Staple a sheet-plastic vapor barrier to the basement side of the walls where condensation is likely to form.

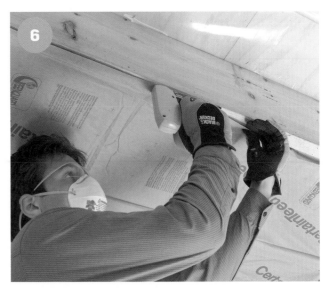

Also insulate the ceiling. Line the joist cavities above the root cellar with sheet plastic before you install the insulation to create a vapor barrier there (the vapor barrier always goes on the warm side of the insulation). Use faced fiberglass batt insulation, or use unfaced fiberglass and install a ceiling covering such as paneling.

Install a wall covering, such as paneling or drywall, over the vapor barrier on the basement side (required for fire resistance). You may cover the wall on the root-cellar side if you wish—there is little point in doing it for aesthetic reasons, but the wallcovering will protect the insulation from damage.

Hang the door. A steel garage service door with a foam core is durable and well insulated. Be sure to install weather stripping around the door to create a seal that minimizes heat transfer.

Remove the basement window sash, if your cellar area has a window. (If not, install a vent in the rim joist—find information on installing a dryer vent for guidance.) Keep the window stop molding in the jambs intact if you can.

Make a ventilation insert panel to replace the window. The panel should have an outflow vent with a manually operated damper so you can regulate the temperature by letting warmer air escape. It should also have an intake vent with ductwork that helps direct cold air down to floor level. On the exterior side, cover the vent openings with insect mesh to prevent rodents and insects from gaining access to your cellar.

Install the ventilation panel insert in the window frame by nailing or screwing it up against the stop molding. Caulk around the edge to prevent insects from getting in. Paint or cover the outside of the panel to weatherproof it.

Provide lighting. If you do not want to install a new, hardwired light and switch (this should be done before walls are covered if you do it), install a stick-up light that operates on battery power. A model with LED bulbs will run for months of intermittent use without a battery change. Add racks and storage features (see next page).

Building a Root Cellar Shelf

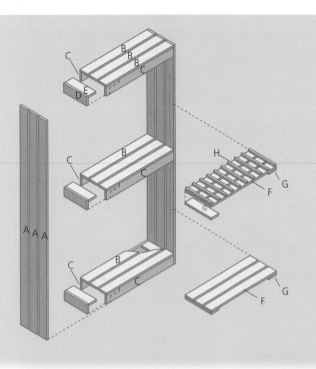

CUTTING MATERIALS

KEY	PART	DIMENSION
A	(6) Side slat	¾ × 3½ × 84" pine
B	(9) Fixed-shelf slat	¾ × 3½ × 30½" pine
C	(6) Fixed-shelf face	¾ × 3½ × 30½" pine
D	(6) Fixed-shelf end	¾ × 3½ × 10½" pine
E	(6) Fixed-shelf stretcher	¾ × 3½ × 10½" pine
F	(6) Adjustable shelf slat	¾ × 3½ × 30⅜" pine
G	(4) Adjustable shelf stretcher	¾ × 3½ × 12" pine
H	(10) Bottle-shelf cleat	¾ × ¾ × 12" pine

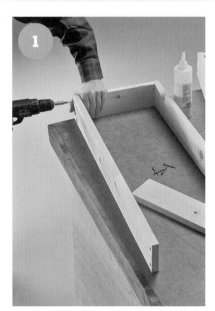

Begin assembling the fixed shelves by cutting the ends and faces to size, and then joining them with glue and counterbored screws. Check with a framing square to make sure the frames you're assembling are square.

Add the stretchers to the tops of the fixed shelf frames. In addition to strengthening the fixed shelf units, the stretchers provide nailing or screwing surfaces for attaching the shelf slats.

Cut the fixed-shelf slats to length, sand them, and attach them to the fixed shelf frames by driving 1¼" screws up through counterbored pilot holes in the stretchers and into the bottom of the slats. Keep your spacing even and make sure the slats do not overhang the frame ends.

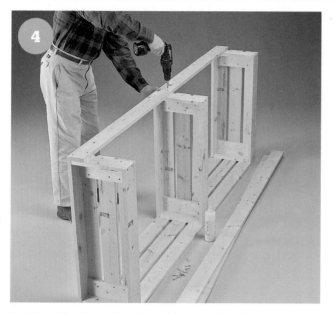

Cut the side slats to length, sand them, and attach them to the outside of the fixed shelf units with glue and counterbored wood screws. Make sure the spacing (¾" between slats) is correct and that all joints are square.

Drill adjustable shelf peg holes in the side slats. To ensure good results, make a drilling template from a piece of perforated hardboard. Use a drill bit the same diameter as your shelf pins, and drill the holes ½" deep. Use masking tape as a drilling depth gauge.

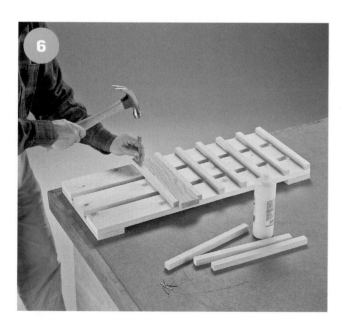

Build an adjustable shelf to support a bottle (wine, for instance) rack, using a 2½"-wide spacer to set the distances separating the shelf cleats. Attach the cleats to the shelf with glue and 4d finish nails. Make the other adjustable shelf.

Insert shelf pins and install the adjustable shelves. Fill screw-hole counterbores with wood plugs, trim flush, and sand. Finish the shelf as desired. Because it is for indoor use, you may leave it unfinished for a rustic look if you prefer.

cider press

This press is the perfect way to make use of the abundant bounty from apple trees on your property. But even if you don't happen to have your own mini orchard, you can make good use of the wide selection of inexpensive apples available at the local co-op or farmer's market (or any place they sell organic apples).

There are few drinks so satisfying as a well-made, home-brewed apple cider, and the process is a simple one. You simply grind the apples into a pulp that you then press to extract the sweet, flavorful juice. If you're only making a small batch, you can chop the apples and process them into pulp in a blender. For larger or multiple loads—if, for instance, you have several trees on your property—you'll want to invest in a full-scale grinder that will make short work of even a large number of apples.

The best ciders incorporate a blend of apples to create an interesting and refreshing flavor profile. Usually, proficient cider makers will include a good amount of very modestly flavored apples (such as Macintosh) as a foundation for the cider, adding tart apples (such as Granny Smiths) to brighten and liven the flavor, and a more flavorful apple to add richness (such as the Orin or Golden Russet). The best way to blend is to crush and press batches of each different apple separately and then blend the ciders so that you can better control the flavor.

When you press and crush the pulp, the cider slowly oozes out of the pulp into the tray below the pressing bucket (a pail in this case). The bucket has to withstand a lot of pressure, and traditionally, press buckets were made from scrap hardwood, with hard staves and hoops and spaces left between the staves for the juice. Because a pressing bucket can be a formidable project in its own right, we've opted for the more convenient 5-gallon plastic pail. These pails are widely available at home centers and hardware stores.

The mash—or crushed apples—is contained within a coarse nylon bag or cheesecloth, which prevents large pieces of apple in the mash from finding their way into the cider. A word of caution about collecting the pressed cider—apple juice can be very acidic, and the acid can react to certain metals, so it's best to catch the juice in a plastic pail and then store it in glass.

A press such as this can process many gallons of delicious cider over a season. It can also be used to crush grapes and extract the juice for homemade wine.

Building a Cider Press

TOOLS & MATERIALS

Cordless drill

Impact driver (optional)

¼" spade bit

½" spade bit

½" metal drill bit

#8 or 9 countersink bit

Miter saw

Jigsaw

Framing square

Clamps

Hydraulic bottle jack (2 ton)

Nylon mesh bag or yard of cheesecloth

Deck screws 1¼", 3"

(2) 7 × ½" hex bolts, washers, nuts

(8) 5" self-tapping lag screws with washer heads

(8) 8 × ¼" self-tapping lag screws with
    washer heads

12 × 16" (or larger) baking pan with
    1 to 1½" high sides

(2) 5-gallon food-safe buckets

Compass

Glue

Sander

Nail set

Eye and ear protection

Work gloves

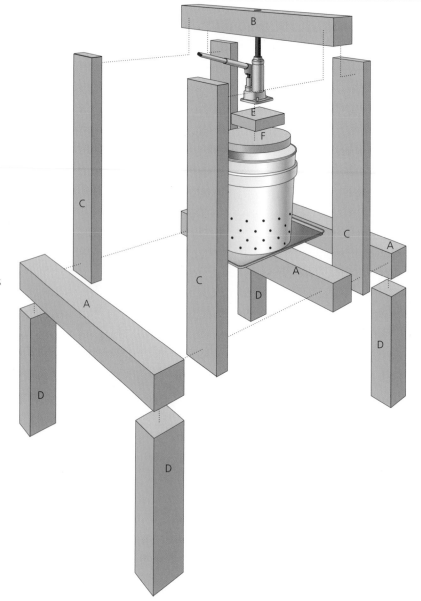

CUTTING LIST

Key	No.	Part	Dimension	Material
A	3	Base boards	3½ × 3½ × 33½"	4 × 4
B	1	Crossbrace	3½ × 3½ × 33½"	4 × 4
C	4	Frame supports	1½ × 5½ × 30"	2 × 6
D	4	Legs	3½ × 3½ × 15"	4 × 4
E	1	Press plate	Bucket dia. × 1½"	1 × 12 (2)
F	1	Press plate backer	1½ × 5½ × 5½"	2 × 6

*Do not use treated lumber for parts that will contact food products.

Lay out the top crossbrace, the center base board, and two frame supports. Make sure the frame is square, then attach frame supports with two 3" screws at each corner. Leave room in the center for bolts. Attach the other two frame supports the same way. Drill holes for the carriage bolts that connect the crossbrace and frame supports, then bolt the frame together.

Align the outside base boards with the center crossbrace. Predrill two ¼" holes at each end of the outside base boards, then fasten the 4 × 4s with two 8" self-tapping lag screws. (If you have an impact driver you can drive the screws in without predrilling.)

(continued)

Building a Cider Press (continued)

Position the press frame on the 4 × 4 legs. With the legs properly aligned, drill two pilot holes at each corner of the outside base 4 × 4s, down into the tops of the legs. Screw the base to the legs using 5" self-tapping lag screws.

Trace the bottom of the bucket on the press plates, then trace an inner line with a compass about ⅛" inside of the line. Cut the line with a jigsaw. Offset the grain direction of the two press plates for more strength, then fasten the two plates together with glue and predrilled, countersunk 1¼" screws. Attach a 5 ½" long 2 × 6 to the center of the plates to distribute the force of the jack.

Sand all the edges of the press plate and check the fit. Also cut several short pieces of 4 × 4 to extend the reach of the jack as it pushes down into the bucket.

Mark three ½" drain holes along the front end of the metal tray, spaced 1 inch apart. Use a nail set to mark the center so the drill bit stays put. Clamp the tray to a piece of scrap wood and drill from the inside out. File away any sharp edges and clean thoroughly.

Drill rows of ¼" holes all the way around the 5-gallon bucket, starting 1" above the bottom and continuing to about halfway up. Scrape and sand the holes smooth and wash the bucket.

Chop, grind, and mash the fruit as fine as you can. Pour it into a mesh bag or cheesecloth in the bucket. The finer the apples are ground up, the more cider you'll produce.

Place the pan on the base with the holes over the collection bucket. Place the jack in the center of the press plate. Place a metal plate or large, thick washer on top of the jack to keep the head from pushing into the wood. Slowly compress the fruit with the jack, adding additional 4 × 4 blocks when the jack is too low in the bucket.

Making the Cider

Once your press is constructed, you're ready to make your first batch of cider. Grind the apples up into mash and place the mash in a nylon mesh press bag (or line the bucket with cheesecloth). Put the bag in the pressing pail and position it so that the press plate is aligned perfectly with the mouth of the pail. You may want to slightly shim the back edges of the back legs, so that the press is tilted forward. This will help the pressed juice run toward the drain holes. Set your collection pail under the drain holes, and pump the jack until juice begins flowing. Continue slowly pumping until no more juice comes out. Now enjoy your first glass of fresh-pressed cider!

219

solar oven

34

There are many effective ways to make a solar cooker—one website devoted to the subject features dozens of photos of different types sent in by people from all around the world. The basic principle is so fundamental that it is easily adapted to a range of styles. We settled on this particular design mostly because it's low-cost to build, working with wood is often easier than manipulating metal, and the unit can cook about any meal you might need to make. However, it's easy enough to modify this design to suit your own food preparation needs.

The cooker is big enough to hold two medium-size pots. All the pieces are cut from one 8-foot 2 × 12 and a sheet of ¾-inch plywood. (The cooker would work just as well with ¼-inch plywood, but we used ¾-inch because it made it simpler to screw the corners and edges together.) The base is made from 1½-inch thick lumber for ease of construction and for the insulation value of the thicker wood, but thinner material would also work.

The foil we used was a type recommended for durability and resistance to UV degradation by an independent research institute. Unfortunately, it was expensive, and if you're just starting out, you may want to do a trial run with heavy-duty aluminum foil. Although foil looks a little dull, it actually reflects solar rays almost as well as specially polished mirrors.

In operation, the cooker is the height of simplicity. The sun's rays reflect off the foil sides and are concentrated at the base of the cooker, where they are absorbed by the black pot. The glass cover (or clear oven cooking bag) helps hold heat and moisture in the pot. The cooker should face the sun. Raise or lower the box depending on the time of year so that you catch the sunlight straight on. Shim the wire rack as needed to keep the pot level.

A solar cooker is an incredibly useful appliance that exploits the limitless energy in sunshine to cook meals large and small.

Types of Solar Cookers

There are many different types of solar cookers. Really, the only requirement is that the sun's rays be captured and focused on whatever is being cooked. Beyond that, the actual construction of the unit can be left to the imagination. However, the most common and popular types of solar cookers can be divided between three groups.

Parabolic Concentrators

Sometimes called curved concentrators, these are a more sophisticated design, but they accommodate only one cooking pot at a time in most cases. However, the shape effectively focuses the sun's rays much better than other types of solar cookers, resulting in higher cooking temperatures. These cook faster, but must be monitored more closely to ensure proper cooking. Crude versions can sometimes be adapted from retired satellite dishes or other parabolic devices.

Box Cookers

These are probably the most popular type of solar cooker because they are so easy to build. The shell can be even be made from found wood or other scavenged materials of odd sizes (which probably accounts for the prevalence of this type of cooker in the third world and impoverished areas). Basically, all you need is a box with a reflective surface inside, and a reflective lid. The box is positioned facing south, and opened at an angle that best directs the sunlight down into the cavity of the box. This type of cooker can be built small or large, is highly portable, and can be constructed to accommodate specifically the dimensions of the cookware that the user already owns.

Panel Cookers

These can be considered a hybrid of the other two styles of cookers. These are a fundamental design that is easy to construct and works well—sometimes better than a box cooker. They are available as kits from self-sufficiency and survivalist manufacturers, but with a little bit of thought and effort, one can easily be constructed from scratch. Be prepared to experiment with the angling of the panels to find exactly the orientation that will work best for your location and situation.

Building a Solar Oven

Key	No.	Part	Dimension	Material
A	2	Base side	1½ × 11¼ × 19"	2 × 2 pine
B	2	Base end	1½ × 11¼ × 16"	2 × 2 pine
C	1	Bottom	¾ × 19 × 19"	¾" ext. grade
D	1	Adjustable leg	¾ × 10 × 17"	¾" ext. grade
E	1	Hood back	¾ × 20 × 33¾"	¾" ext. grade
F	1	Hood front	¾ × 10 × 25¼"	¾" ext. grade
G	2	Hood side	¾ × 20 × 31¼"	¾" ext. grade
H	1	Lens	¼ × 17¼ × 17¼"	Tempered glass

TOOLS & MATERIALS

Straightedge

Circular saw

Jigsaw or plunge router

Tape measure

Drill/driver with bits

Speed square

Stapler

#8 countersink bit

¾" × 4 × 8' BC or better plywood

2 × 12 × 8' SPF SolaReflex foil or heavy-duty
  aluminum foil

Bar clamps

1⅝", 2½" deck screws

Clear silicone caulk

Mid-size black metal pot with glass top

Wire rack

No-bore glass lid pulls

¼ × 2" hanger bolts with large fender
  washers and wingnuts

Sander

Glue

Eye and ear protection

Work gloves

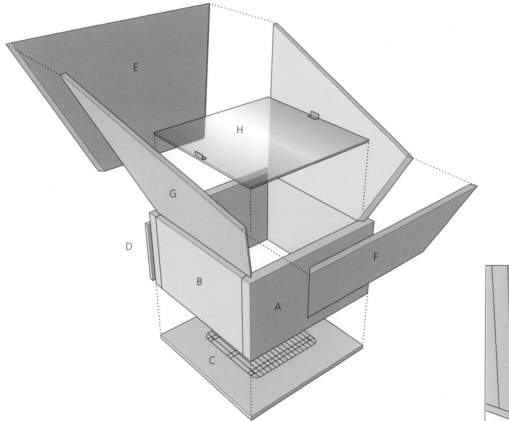

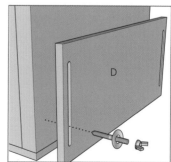

Building a Solar Oven

Cut the four 2 × 12 base pieces to length according to the cutting list. Arrange the base parts on a flat work surface and clamp them together in the correct orientation. Check for square and adjust as needed. Drill pilot holes and fasten the pieces together with 2½" deck screws.

Lay a sheet of plywood on the work surface with the better side facing up. Mark and cut the bottom first. Rest the full sheet of plywood on a couple of old 2 × 4s.

To create the panels that form the reflector, you'll need to make beveled cuts on the bottom and sides so the panels fit together squarely. Mark two 20 × 76" long pieces, measuring from the two factory edges so the waste will be in the middle. Set your circular saw base to 22½°, then cut along the line you drew at 20". Cut the other piece starting from the opposite end of the plywood. You should end up with two mirror-image pieces.

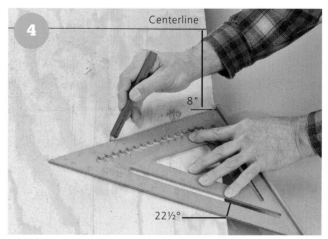

Reset your saw base to 0°, then cut each 20"-wide panel in half so you have four 20 × 38" panels, each with one beveled 38" edge. With the beveled edge facing up and closest to you, draw a centerline at 18" on each panel, then make marks on the beveled edges at 8" on both sides of the centerline. Position a speed square so it pivots at the 8" mark, then rotate the speed square away from the centerline until the 22½° mark on the speed square meets the top of the beveled edge. Draw a line and use a straightedge to extend the line to the other edge (the factory edge) of the plywood. Repeat at the other 8" mark, flipping the speed square and rotating it away from the centerline so the lines create a flat-topped triangle. Set the base of your circular saw at 40°, then cut along the angled lines. Mark and cut the remaining three panels in the same fashion.

Finish cutting the reflector parts to final size and shape. **TIP:** After you've laid out your cutting lines, set the workpiece onto a pair of old 2 × 4s. Tack the workpieces to the 2 × 4s with finish nails, driven into the waste area of the panels. Keep the nails at least a couple of inches from any cutting line. Set your saw so the cutting depth is about ¼" more than the thickness of the workpiece and then make your cuts.

Assemble the reflector. Brace two of the reflector sides against a square piece of scrap plywood clamped to the work surface, then join the edges with screws driven into countersunk pilot holes. Repeat for the other two pieces. Join the two halves together with four screws at each corner, completing the reflector. The bottom edges should be aligned. The top edges won't match perfectly, so sand them smooth.

Compound Miter Corner Cuts

The sides of this solar cooker box are cut with the same basic technique used to cut crown molding. Instead of angling the crown against the miter saw fence in the same position it will be against the ceiling—a simple 45° cut that is easy to visualize—you have to make the compound cuts with the wood lying flat, which makes it mind-bendingly difficult to visualize the cut angles. For the dimensions of this cooker, a 40° bevel cut along the 22½° line will form a square corner. If you change the 22½° angle, the saw cut will also change.

If you remember your geometry, you can work all this out on paper. But bevel guides on circular saws are not very precise, and 40° on one saw might be more like 39° on a different brand; test cuts are the best way to get the angle right. Make the first cuts a little long and then try them out.

The easiest way to avoid a miscut is to lay all the pieces out with the bases lined up and the good side of the plywood up. Mark the 22½° lines for the sides, then cut the 40° angles on one edge of all four pieces. Next, flip the piece around and cut the 40° angle on the other side. Remember, the 40° cut should angle outwards from the good side of the plywood, and the pieces should all be mirror images.

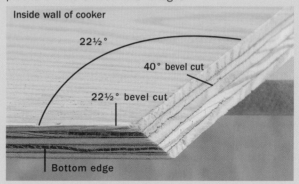

(continued)

Building a Solar Oven (continued)

Cut the adjustable leg with parallel slots so the leg can move up and down over a pair of hanger bolts, angling the cooker as necessary. Outline the slots so they are ⅜" wide (or slightly wider than your hanger bolt shafts). Locate a slot 2" from each edge of the adjustable leg. The slots should stop and start 2" from the top and bottom edges. Cut the slots with a jigsaw or a plunge router.

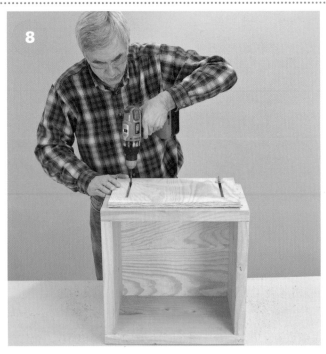

Screw the plywood bottom to the base. Set the adjustable leg against one side of the base, then drill guide holes and install the hanger bolts to align with the slots. Center the bolts at the same height: roughly 2½" up from the bottom of the box. Use large fender washers and wing nuts to lock the adjustable leg in position.

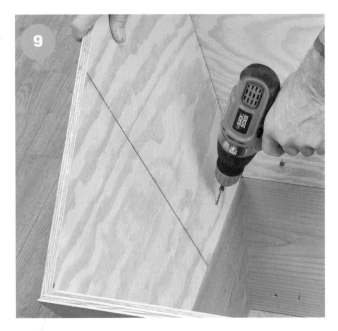

Fasten the reflector to the base with countersunk 2½" deck screws. Angle the drill bit slightly as you drill to avoid breaking the plywood edge. Use two screws per side.

Cut pieces of reflective sheeting to fit the sides of the reflector as well as the base. You can use heavy-duty aluminum foil, but for a sturdier option try solar foil. Cut the pieces large enough to overlap at the edges.

Take measurements to double-check the glass lid size. Ideally, the lid will rest about 1" above the top opening of the box. Order glass with polished edges. You can also use a clear plastic oven bag instead of the glass. Either will trap heat and speed up the cooking.

Glue the reflective sheeting inside the base and reflector, overlapping the sheets. Use contact cement or silicone caulk and staple the edges to reinforce the glue (use diluted white glue with a paintbrush instead of contact cement if you're using aluminum foil.) Smooth out the reflective material as much as possible; the smoother the surface is, the better it will reflect light.

Getting a Handle on Glass

Because it is virtually impossible to lift the glass lid from above, you'll need to install handles or pulls designed to attach to glass (available from woodworking hardware suppliers). The simplest of these require no drilling. You squeeze a bead of clear, 100 percent silicone into the U-channel of the lid handle, then slide the handle over the edge of the glass.

Caulk the joint between the angled top and the base with clear silicone caulk. Set a wire rack inside the oven to keep the cooking pot slightly elevated and allow airflow beneath it.

homestead amenities

The most important word in homestead is "home." After all, everything you're doing in a move toward self-sufficiency is aimed at making your home a better place to live. You can actually consider homesteading the improvement of your many homes—the world, your region, and the plot of land that holds that structure where you lay your head at night. This section is, in the final analysis, all about that last one.

As anyone knows, a home can either be warmly inviting, comforting and wonderful, or cold and off-putting, and never quite settled. The amenities in this chapter focus on moving your home toward the former rather than the latter. You see, when you make your home—including the inside and outside—more functional, you make it a nicer place to live.

You can consider these practical amenities. Having a handy place to stack and season your firewood, a countertop wine rack, or a solar heat panel is all about convenience. Projects like these also often provide viable alternatives to using expensive appliances that are energy hogs. Do a load of laundry in a manual laundry machine and you save electricity even while you discover the meditative bliss of losing yourself in 20 minutes of non-taxing labor performing an essential household task. Beyond convenience and the other rewards of these projects lies something else. Something that isn't quite as tangible, but to anybody who values self-sufficiency and a return to a simple life, is just as important. That is handmade craftsmanship.

Surrounding ourselves with conveniences isn't hard. A few hours spent at an appliance store can do that. But surrounding ourselves with conveniences that give us a connection to our home and remind us that yes, we can do for ourselves . . . well, those are simply the best type of conveniences and true homestead amenities.

firewood shelter

35

Everyone knows that wood burns best when it's dry. But properly dried, or "seasoned," firewood isn't just about making fires easy to start and keep burning. Seasoned wood burns hotter and cleaner than unseasoned ("green") firewood, resulting in more heat for your home, reduced creosote buildup in your chimney, and lower levels of smoke pollution going into the air.

Seasoning freshly split wood takes at least six months in most areas, but the longer you can dry it the better. The best plan is to buy firewood (or cut and split your own) as early as possible and stack it in a well-ventilated shelter with a good roof. This shelter will keep your wood covered through snow and rain while providing ample ventilation and easy access to the stack. It also looks better than any prefab shelter and is easy to modify with different materials or dimensions.

An optional feature is a storage locker at either end of the structure, perfect for storing your axe, gloves, and other tools and supplies. A simple bin or locker at the other end can hold kindling or provide more protected storage space. The roof trusses add some custom detailing and actually simplify the construction. You can top the roof with fiberglass panels, as shown, or any other type of roofing material.

The shelter is sized to hold a half-cord of split firewood cut to 16-inch lengths, and stacked two deep. If you use shorter logs, you can stack them two or three deep. For longer logs or to accommodate a whole cord of wood, you can easily modify the shelter dimensions to fit (see Resizing Your Shelter, right).

Resizing Your Shelter

A cord of split and stacked firewood measures 4 × 4 × 8', or 128 cubic feet Standard log lengths are 12, 16, and 24", with plenty of variation in between (they're logs, after all, not trimwork). Sizing your shelter depends on the length of your logs, how much wood you want to store, how you want to stack it (two-deep, three-deep, etc.), and how much roof coverage you'd like extending over the sides of the stack.

Modifying this shelter design is simple. First determine the floor platform size and the height between the floor and the roof—these dimensions give you the overall wood storage capacity. The rest of the structure can be sized and built as you go, as all of the other elements are based on the spans between the corner posts. For a significantly longer shelter design, add full-height intermediate posts and additional trusses to strengthen the roof assembly; for a much deeper (more square-shaped) structure, use intermediate posts and/or larger lumber for the trusses and roof beams.

A shelter like this is a handsome addition to a yard or alongside the house and will keep your firewood nice and dry as it ages. Add storage lockers at either end of the shelter to make it even more useful.

231

Building a Firewood Shelter

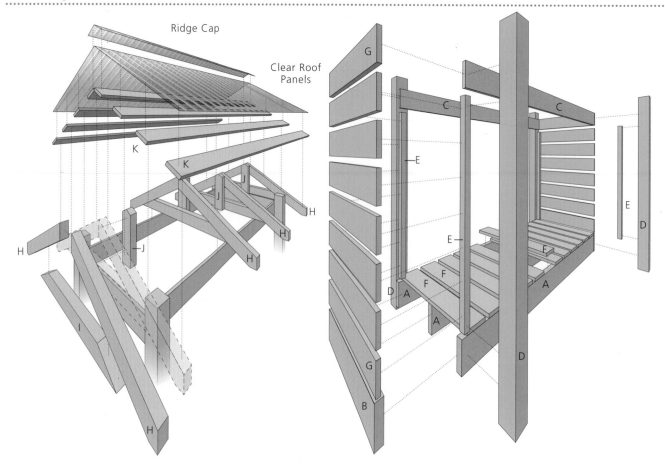

Ridge Cap

Clear Roof Panels

CUTTING LIST

Key	No.	Part	Dimension	Material*
A	3	Floor beams	1½ × 7¼ × 81½"	2 × 8
B	2	End beam	1½ × 7¼ × 26"	2 × 8
C	2	Roof beam	1½ × 5½ × 81½"	2 × 6
D	4	Corner post	3½ × 3½ × 67"	4 × 4
E	4	Corner cleat	1½ × 1½ × 48	2 × 2
F	12	Deck board	¾ × 5½ × 28"	1 × 6
G	16	Siding	¾ × 5½ × 28"	1 × 6
H	8	Top chords	1½ × 3½ × 30"	2 × 4
I	4	Bottom chords	1½ × 3½ × 36"	2 × 4
J	4	Truss center post	1½ × 3½ × 14½"	2 × 4
K	8	Purlins	¾ × 3½ × 92"	1 × 4

* All lumber should be pressure-treated; posts and floor beams should be rated for ground contact. All hardware must be corrosion-resistant for outdoor exposure.

TOOLS & MATERIALS

Hammer
Tape measure
Carpenter's square
4' level
Circular saw with wood- and metal-cutting blades
Reciprocating saw or handsaw
Cordless drill and bits
Clamps
Socket wrench
Miter saw (optional)
8d galvanized siding nails
Deck screws 2", 2½", 3½"
(16) 6 × ⅜" carriage bolts with washers and nuts
⅞" roofing screws with neoprene washers
Joist hangers (for 2 × 8) with recommended fasteners (2)
Joist hanger nails
Rafter ties
16d galvanized common nails
Fiberglass roofing panels
Speed square
Eye and ear protection
Work gloves

Assemble the frame with 3½" deck screws driven through the side beams and into the ends of the end beams. Install the floor center beam using joist hangers and joist hanger nails or screws. Measure from diagonal to diagonal in both directions to make sure the frame is square, then temporarily screw a board across two corners to hold it in place. Set the floor frame on a flat, level surface, on 1"-thick spacers.

Position the posts so they extend ¾" beyond the ends, to leave room for the 1 × 6 slats. Clamp each corner post to the outside of the frame, check for plumb and square, and then screw it in position with two screws through the 2 × 8 (remember to leave room for bolts). Drill two ⅜" holes through the posts and side beams, angling the drill so the bolts go from the middle of the 4 × 4 to the open part of the 2 × 8. Anchor each post with two carriage bolts.

Mark the center line across the floor beams, then position and screw down the deck boards, starting ¾" to each side of the center line. Space the boards 1⅜" to 1½" apart and fasten with 2½" screws. Overhang the beams about ½" on each side. Check and adjust the spacing as you get close to the ends—pressure-treated wood can vary as much as ⅛" in width, depending on how dry it is.

(continued)

Building a Firewood Shelter (continued)

Clamp each roof beam to the opposing corner posts so that the top edge of the beam is 3½" below the top ends of the posts, and the ends of the beams are ¾" in from the outside faces of the posts. Hold the beams in place with a 2½" deck screw at each end, then drill holes and fasten the beams to the posts with ⅜" carriage bolts.

Enclose the openings on the ends of the shelter with 1 × 6 slats secured to 2 × 2 cleats mounted to the inside faces of the posts. Install the cleats ¾" back from the outside faces of the posts so that the slats are flush with the post faces. Space the slats roughly 1¾" apart.

Miter the end cuts on the top chords at 30° on the miter saw. Mark the ends of the bottom chord at 60° with a speed square, then cut with a circular saw. Hold the center post of the truss in place and mark it for cutting. Build the trusses on a flat surface, fastening the top chords to each other and to the bottom chord with 3½" screws. Use 2½" screws to fasten the center post.

Mark the outside faces of the corner posts 3½" down from their top ends. Clamp the two outer trusses to the posts with the bottoms of the trusses on the 3½" marks, then fasten them with 16d galvanized nails.

Position the remaining two trusses on top of the roof beams, spacing them evenly, and screw them to the roof beams using rafter ties.

Nail the purlins across the rafters on both sides with 8d nails. Make sure the rafters are spaced evenly before you nail. The ends of the purlins should extend 3" beyond the outside trusses.

Cut sections of clear roofing sheets to the desired overhang. Clamp the sheets to a solid base and cut with a thin-kerf fine-cutting carbide blade (or a ferrous metal-cutting blade for metal). Sand or file sharp or jagged edges. Fasten the roofing to the purlins using roofing screws with neoprene washers. Overlap the sheets and screw through both pieces along the joint. Install ridge caps using the same screws.

frame loom

36

Weaving your own textiles can be incredibly relaxing, enjoyable, and fulfilling. It's a way to reclaim a heritage craft and create beautiful fabric pieces, from scarves to rugs. The trick is to build your skills on a small loom first, and then progress to a larger, more complicated loom. This naturally means starting small, but the idea is to build on your successes, and avoid the frustration that can come from handling a more sophisticated apparatus before you're ready.

Frame looms like the one in this project are ideal for the beginner or intermediate weaver. They are easy to handle, portable, and small enough that they don't take up much room. Just the same, you'll be able to create decorative fabric panels for hanging, small runner rugs, scarves, and even panels that can be sewn together to create more involved projects like a quilt.

The construction of this loom is simple and straightforward. The example here is a fairly standard size, but don't be afraid to resize the dimensions to suit your own needs and preferences. Just be careful not to make it too big or the frame will have a tendency to flex as you work, making the weaving more difficult. We've also included legs on this frame loom to make weaving more comfortable. Adjust the position so that you can sit comfortably and weave without excessive reaching or fumbling.

As the name implies, you begin by building a fundamental frame to which the supporting (or "warp") fibers will be secured. The "weft" threads that run horizontally are then woven through these warp threads. This is the basic process of any loom—the technique just becomes more involved the bigger and more complex the loom. The terms associated with loom weaving can be a little confusing to the beginner, so we've included a glossary to keep things straight. No matter what words you use, however, the result will be a fabric that you've created with your own hands—no fabric store or mill necessary!

Create marvelous handicrafts with a simple frame loom like this one. It's easy to use and will help develop your weaving skills, should you ever want to step up to a standalone loom and bigger textiles like rugs and bedspreads.

237

Building a Frame Loom

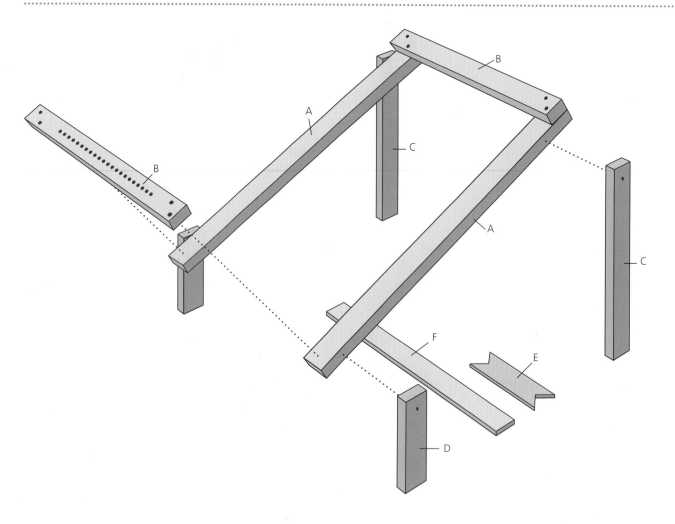

CUTTING LIST

Key	No.	Part	Dimension	Material
A	2	Frame sides	¾ × 1½ × 24"	1 × 2
B	2	Frame ends	¾ × 1½ × 18"	1 × 2
C	2	Back legs	¾ × 1½ × 12"	1 × 2
D	2	Front legs	¾ × 1½ × 5"	1 × 2
E	1	Shuttle	¼ × 1½ × 6"	Mull strip*
F	1	Shed stick	¼ × 1½ × 18"	Mull strip

* Can also be made from a paint stir stick.

TOOLS & MATERIALS

Measuring tape
Cordless drill and bits
Handsaw or utility knife
1½" finish nails
2½" machine screws and matching wing nuts (10–32)

Rubber non-slip furniture leg pads
Construction adhesive
Eye and ear protection
Work gloves

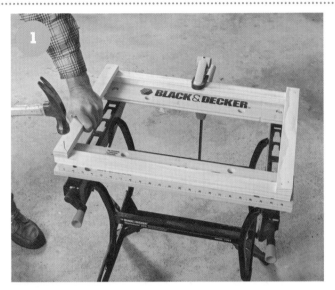

Set the frame sides flat on a clean, level work surface and align the frame ends across either end of the sides. Drill pilot holes and nail the ends to the sides using two finish nails at each corner. Check for square as you work and adjust as necessary.

Mark and drill holes through the tops of the four legs and through each frame side. The legs are attached 2" in from each end of the frame. The holes need to be the same diameter or just slightly bigger than the machine screws you've selected. After the holes are drilled and tested, mark the legs for each side and remove the bolts.

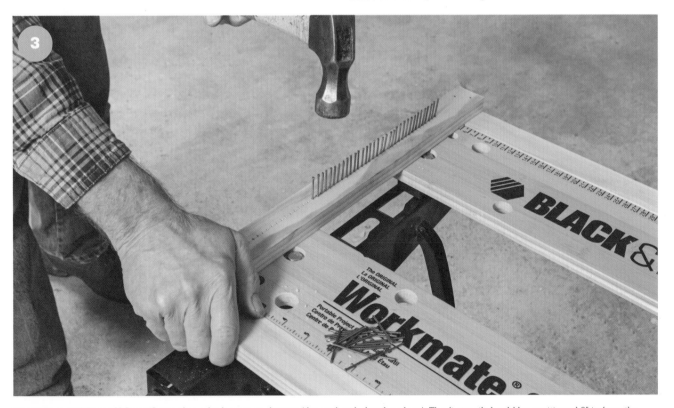

Mark the guide nail locations along the bottom end piece (the end with the short legs). The first nail should be positioned 3" in from the end, with nails every ¼" along the face of the end piece. The last nail will be located 3" in from the opposite end. Mark all the locations, then predrill the holes so the wood doesn't split. Drive a finish nail at each point, halfway into the wood.

(continued)

Building a Frame Loom (continued)

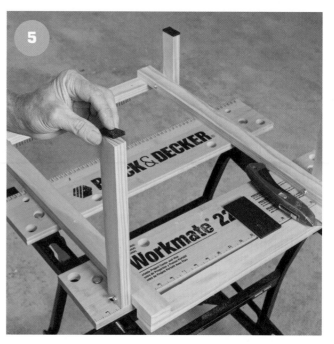

Drill a ⁵⁄₃₂" hole in the shuttle face, ¾" from the end, and centered between the edges. Repeat on the opposite side. Mark lines from the corners to the hole, and use a jigsaw or handsaw to cut a triangle from the end of the shuttle. Repeat on the opposite side.

Attach the legs to the loom frame with the machine screws and wing nuts. Cut the rubber pads to the fit the bottom of the legs and glue them in place with construction adhesive. These will keep the frame from sliding when you are weaving.

Weaving Terms

There's a whole language around working with a loom. Once you learn them, the terms are pretty common sense, and they're necessary if you ever graduate to a more complex freestanding loom.

Warp: This is the main set of threads or yarn fibers that run vertically up the loom. The word is also used to describe the process of attaching these threads.

Weft: The yarn or fibers that run horizontally between the warp threads.

Heddle: To make things easier, avid weavers use this tool to maintain a space between alternating warp threads. This makes it much easier to weave the weft in and out of the warp. The project described here uses a shed stick for the same function.

Shuttle: You can weave the weft by hand, or do what most weavers do and wrap the yarn or fiber around this handy tool. It's basically just a flat stick with grooves or cutouts, around which the yarn is wrapped. The shuttle is then passed through the warp to quickly create the weave.

Shed: The space you create with a heddle or spacer stick—through which the shuttle and weft yarn passes—is called a shed.

Using Your Loom

The basic idea behind this loom—and any loom for that matter—is to interweave yarn in perpendicular directions to create strong textile panels. Although yarn is the most commonly used fiber, other fibers can be used for different purposes, such as weaving placements for a dining table. In any case, begin by tying one end of the yarn around the bottom end (the end with the nails) of the frame. Secure it with a double slip knot and wind the warp under the top end, over the top end, then under the bottom end and back over, so that the warp looks like a figure 8. Use the brads as guides to create regular spacing between the strands of yarn. Slide the shed stick between the warp near the top end of the loom, then move it down toward the bottom end and turn it on edge, creating a space for the shuttle to pass through. Tie the end of the weft to the side of the frame at the bottom, then pass the shuttle through the

warp and tamp the weft down to the bottom. For the next weft, slide the shed stick from the other side, over and under the opposite warps. In other words, where the weft went under a warp, it should now go over it. This can be painstaking work if the warp goes through every nail, but can go much quicker if the warp is spaced more widely. After the shed stick is all the way through, turn it on edge again and pass the shuttle back through. Then pull the weft tight, push it to the bottom, and repeat the first step. Work back and forth to create the weave, tamping down each weft row with the shed stick, so that it's snug to the row beneath it. (There are specialized tools called "beaters" for this purpose, but on a small loom like this one, the shed stick works just as well.) When you're done, untie or cut the warp at either end, and either leave the loose strands, trim them, or finish the piece with a fabric edge band.

Shed stick

Shuttle

solar still

37

With local water supplies increasingly under threat from contaminants such as the runoff from large agricultural operations, ensuring a safe water supply is more of a concern than ever— especially for homeowners who draw their water from a local well. This still can be a great backup, and one more way to ensure self-sufficiency.

The box is built from ¾-inch BC-grade plywood, painted black on the inside to absorb heat. We used a double layer of plywood on the sides to resist warping and to help insulate the box, with an insulated door at the back and a sheet of glass on top. As the box heats up, the water evaporates and heavier contaminant in the vapor is left behind.

Finding the right lining to hold the water inside the box as it heats and evaporates can be a challenge. The combination of high heat and contaminants can corrode metals faster than usual and cause plastic containers to break down or off-gas, imparting an unpleasant taste to the distilled water. The best liners are glass or stainless steel, although you can also coat the inside of the box with two or three coats of black silicone caulk (look for a type approved for use with food). Spread the caulk around the bottom and sides with a taping knife. After it dries and cures thoroughly, just pour water in—the silicone is impervious to the heat and water.

We chose to paint the inside black and use two large glass baking pans to hold the water. Glass baking pans are a safe, inexpensive container for dirty or salty water, and they can easily be removed for cleaning. We used two 10 × 15-inch pans, which hold up to 8 quarts of water when full. To increase the capacity, just increase the size of the wooden box and add more pans.

The operation of the distiller is simple. As the temperature inside the box rises, water in the pans heats up and evaporates, rising up to condense on the angled glass, where it slowly runs down to the collector tube and then out to a container.

The runoff tube is made from 1-inch PEX tubing. Stainless steel can also be used. However, use caution with other materials—if in doubt, boil a piece of the material in tap water for 10 minutes, then taste the water after it cools to see if it added any flavor. If it did, don't use it.

A solar still is a very simple device for ensuring clean, drinkable water—courtesy of the sun.

Building a Solar Still

CUTTING LIST

Key	No.	Part	Dimension	Material
A	1	Base liner	¾ × 23¾ × 19"	Rigid insulation
B	1	Base	¾ × 23¾ × 19"	Plywood
C	1	Inner front frame	¾ × 5¾ × 19"	Plywood
D	1	Outer front frame	¾ × 5⅝ × 20½"	Plywood
E	2	Base	1½ × 3½ × 22½"	2 × 4
F	1	Inner rear frame	¾ × 3 × 20½"	Plywood
G	1	Door inner	¾ × 5⅞ × 20½"	Plywood
H	1	Door outer	¾ × 9 × 20½"	Plywood
I	2	Outer side frame	¾ × 9⅛ × 5⅛ × 26¾"	Plywood
J	2	Inner side frame	¾ × 8⅞ × 5⅝ × 24½"	Plywood
K	1	Window	27¼ × 22 × ⅛"	Tempered glass
L	1	Drain tube	1"	PEX tubing, cut to length

TOOLS & MATERIALS

Drill/driver with bits
Circular saw
Straightedge
Caulk gun
Utility knife
Clamps
Tape measure

(1) ¾" × 4 × 8' sheet of
    BC exterior plywood
(2) 1½" galvanized hinges
Self-adhesive weatherseal (8')
Insulation
Knob or drawer pull
27¼ × 22 × ⅛"

(minimum) glass
Silicone caulk
High-temperature black paint
1" PEX tubing
(2) 10 × 15" glass baking pans
Wood glue
1¼", 2", 2½" deck screws

Painter's tape
Sliding bolt catch
Reflective foil
Eye and ear protection
Work gloves

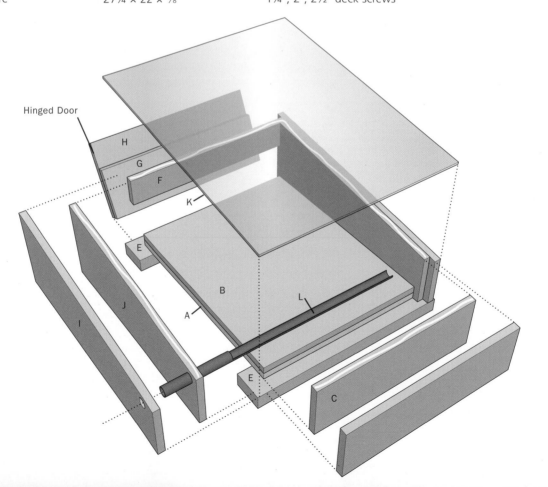

Hinged Door

Mark and cut the plywood pieces according to the cutting list. Cut the angled end pieces with a circular saw or tablesaw set to a 9° angle.

Cut the insulation the same size as the plywood base, then screw both to the 2 × 4 supports with 2½" screws.

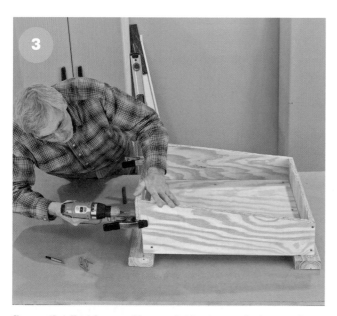

Screw the first layer of front and side pieces to the base and to each other, then add the back piece. Predrill the screws with a countersink bit.

Glue and screw the remaining front and side pieces on, using clamps to hold them together as you predrill and screw. Use 1¼" screws to laminate the pieces together and 2" screws to join the corners.

(continued)

Building a Solar Still (continued)

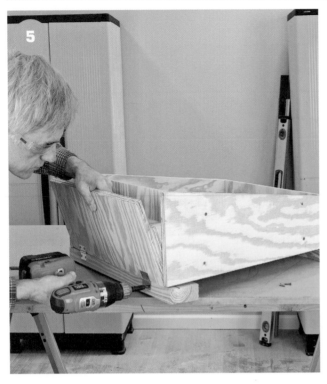

Glue and screw the hinged door pieces together, aligning the bottom and side edges, then set the door in position and screw on the hinges. Add a pull or knob at the center.

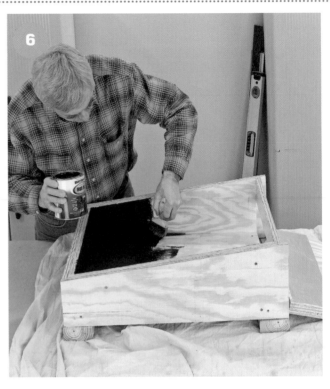

Paint the inside of the box with black high-temperature paint. Cover the back and the door with reflective foil glued with contact cement. Let the paint dry for several days so that all the solvents evaporate.

Apply weatherseal around the edges of the hinged door to make the door airtight.

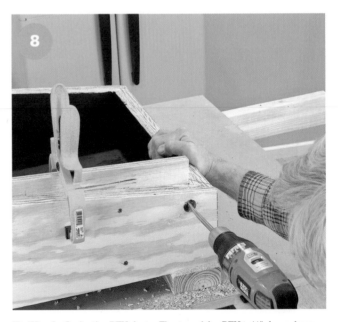

Drill a hole for the PEX drain. The top of the PEX is ½" down from the top edge. Clamp a scrap piece to the inside so the drill bit doesn't splinter the wood when it goes through.

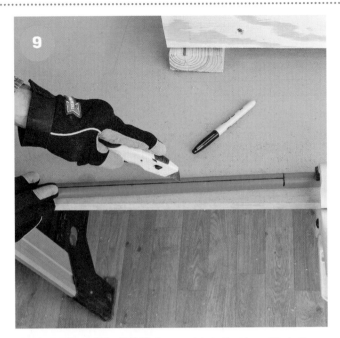

Mark the first 19" of PEX, then cut it in half with a utility knife. Score it lightly at first to establish the cut lines.

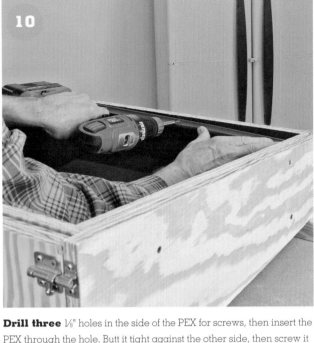

Drill three 1/8" holes in the side of the PEX for screws, then insert the PEX through the hole. Butt it tight against the other side, then screw it in place, sloping it about 1¼".

Wipe a thick bead of silicone caulk along the top edge of the PEX to seal it against the plywood.

Shim the box level and tack a temporary stop to the top edge to make it easy to place the glass without smearing the caulk. Spread a generous bead of caulk on all the edges, then lay the glass in place. Tape it down around the edges with painter's tape, then let it set up overnight.

manual laundry washer

38

Today's high-efficiency washers and driers are marvels of modern technology. They use less electricity and water than their predecessors and include more features than ever. However, for all their improvement on past models, they still use a significant amount of energy (especially when someone does a small, "quick" load—you know who you are), as well as a good deal of precious water.

What's more, washing clothes hasn't changed. The whole idea remains astoundingly simple: agitate the garments in soapy water for a period of time until dirt loosens and releases from the clothes. That basic function doesn't necessarily need a high-tech solution. If you're looking for something more sustainable, more environmentally friendly, and cheaper, you've found it—a simple, manual laundry machine.

This unit is not a challenge to build. All you need are a few basic carpentry tools, a few pieces of hardware, and a couple hours to assemble it. You can build it from components you'll find at any well-stocked hardware store or large home center. The capacity probably won't rival the drum for your current washer, but is certainly large enough to handle most loads of laundry or animal blankets.

Once you've put your washer together, actually doing a load of laundry entails plunging the handle repeatedly for about 20 minutes. The machine makes great use of the lever principle so anyone—even someone with moderate strength and stamina—can easily wash a load of laundry by hand. Use biodegradable soap and you can merely empty the bucket in your landscaping as gray water. Hang the clothes up to dry (use the Clothesline Trellis on page 122) and you're done!

Save electricity, get a little exercise, and conserve water with this handy manual clothes washer. The ingenious lever action ensures that clothes are properly agitated and thoroughly washed.

Building a Manual Laundry Washer

TOOLS & MATERIALS

Cordless drill and bits
Table saw or circular saw
Jigsaw
5-gallon plastic bucket with lid
(4) 2½ × ⅝" corner braces
(1) ¼ × 6" machine screw and nut or wing nut
(1) ¼ × 4" machine screw and nut or wing nut
Deck screws 2½", 3", 3½"
¾" stainless-steel wood screws for corner braces
¾" stainless-steel bolts, nuts and washers for pail lid agitator
Large bucket or tub
Clamp
Chisel
Sander
Utility knife
Eye and ear protection
Work gloves

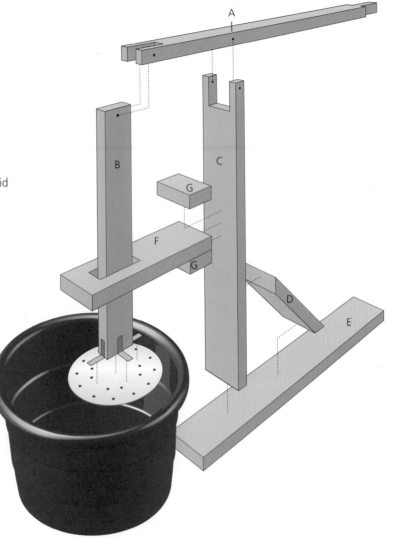

CUTTING LIST

Key	No.	Part	Dimension	Material
A	1	Handle	1½ × 3½ × 51"	2 × 4
B	1	Plunger	1½ × ½ × 25"	2 × 4
C	1	Support	1½ × 5½ × 36"	2 × 6
D	1	Brace	1½ × 5½ × 15"	2 × 6
E	1	Base	1½ × 5½ × 40"	2 × 6
F	1	Guide	1½ × 5½ × 18"	2 × 6
G	2	Cleats	1½ × 3½ × 5½"	2 × 4

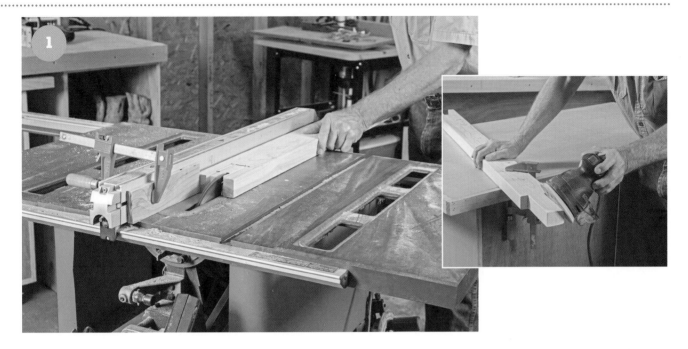

Cut the slot in the end of the handle to accept the plunger by making multiple passes with the table saw. Clamp a stop to the fence so that all the cuts are the same length. The slot should be cut centered on the face of the handle, 1½" wide by 4" deep. Clean the slot out with a chisel after cutting it and sand smooth. Cut and sand the opposite end of the handle to make a grip that's 1½" wide by 5" long. (If you don't have a table saw, just cut the slot with a jigsaw and square the end with a chisel.)

Cut the plunger to length. Mark a hole for the plunger at 1¾" from the side and 1" from the end so that it lines up with the end of the handle and extends ¼" beyond the top of the handle. Drill a ¼" hole centered in the side of the handle slot, 1¾" from the end. Continue the hole through the plunger and opposite slot. Use a drill press if you have one to get a straight hole; otherwise, mark and drill the holes in both sides of the handle, run the bit all the way through to straighten the hole, then drill through from alternate sides of the handle through the plunger. (If the hole is angled too much it may cause the plunger to bind when you move it up and down. If this happens, just redrill the holes with a ⁵⁄₁₆" bit.)

Mark and cut out the slot in the end of the handle support board. The slot should be centered on the face of the support, 3½" wide by 3" deep. Drill the ¼" pivot hole through the sides of the slot ¾" from the end and ¾" from the side. Drill the matching hole in the handle at 13¾" from the center of the hole plunger hole.

(continued)

Building a Manual Laundry Washer (continued)

Attach the base to the opposite end of the support with 3½" deck screws. Miter the ends of the brace 45° and fasten it to the base and support with toenailed 2½" deck screws.

Mark the guide hole for the plunger arm in one end of the guide board. The outer edge of the hole should be 2½" from the end of the board. The hole will be a rectangle, 2 × 5½", with the long sides parallel to the long sides of the guide. Center it on the face of the guide board and drill holes at all four corners. Use a jigsaw to cut out the hole.

Screw the guide to the support, on the opposite side from the brace, using 3½" deck screws. It should be positioned 18" up from the base. Screw one cleat above and one cleat below the guide, snug to it, using 3½" deck screws. Also screw the support and braces together for more rigidity.

Attach the handle to the support with a ¼ × 6" machine screw and bolt or wing nut. Use washers on either side. The bolt should slide through the hole in the handle fairly easily. If it doesn't, run the drill bit through a few more times.

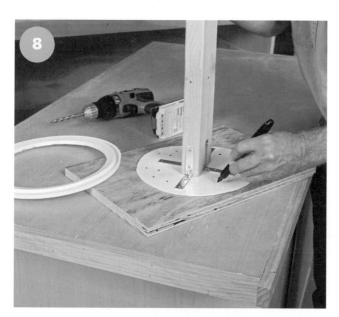

Attach 2½ × ⅝" stainless-steel corner braces to the sides and faces at the bottom end of the plunger. Screw the braces in place with ¾" stainless steel wood screws. Use a utility knife to cut the pail lid along the inside seam, then remove the lip. Center the end of the plunger on top of the lid and mark the location of the holes in the corner braces. Drill the holes for the corner braces, and then drill a varied pattern of about 20 additional ⅜" holes spaced evenly around the lid. This will allow water to pass through when you are agitating the laundry.

Attach the pail lid to the end of the plunger with ¾" stainless-steel bolts, washers and nuts. Slide the plunger arm up through the guide hole in the guide. Secure the tongue at the end of the plunger in the handle groove using a 4" machine screw, a bolt or wing nut, and washers. Check the operation of the handle and plunger and adjust as necessary. You can use a 5-gallon bucket for washing small loads or a larger tub for dropcloths, horse blankets, and other large items.

solar heat

39

Most homes, and especially older ones, are beset by one stubborn room that just never quite warms up—whether it's a remote second floor bedroom, a kitchen or dining room with large, north-facing windows, or a main floor office in the corner closed off from the home's central heating system. Operating an electric space heater can help you warm up in the short-term, but is not an energy-efficient or long-term solution to this problem. One very efficient long-term solution, however, is to build and install hot-air solar panels. Even if you don't have a cold spot, a supplementary solar heat system can save plenty of energy dollars.

Using solar energy to heat a cold space in your home is a great way to harvest the sun's energy and supplement your home's heat in these problem areas. Solar hot air panels are fundamentally different than photovoltaic panels—this style is designed to use the sun's energy to heat the air inside each box rather than to create electricity. Mounted on a south-facing wall or on the roof, solar hot air panels collect air from inside your home and blow or draw it through the thermal solar panels, which are essentially a series of metal ducts in a black box under tempered glass. As the air moves through the ductwork, the sun's rays cause it to heat to high temperatures. Then, at the end of the duct, another vent moves the air back into your home's heating ductwork or an interior vent, sending the now-heated air right into the home.

When combined, these three DIY "hotboxes" introduce enough hot air into this home to carry 30 to 40 percent of the home heating load.

You can build solar hot air panels yourself. This style seen here is simple: essentially, a box, a series of ducts, and a piece of glass. The panels are permanently installed and ducted into your home, complete with automated thermostatic controls. In this project, we'll walk you through one version of a solar hot air panel designed and installed by Applied Energy Innovations of Minneapolis, Minnesota (see Resources page 348), with the help of homeowner Scott Travis.

Anatomy of a Hot Air Solar Panel

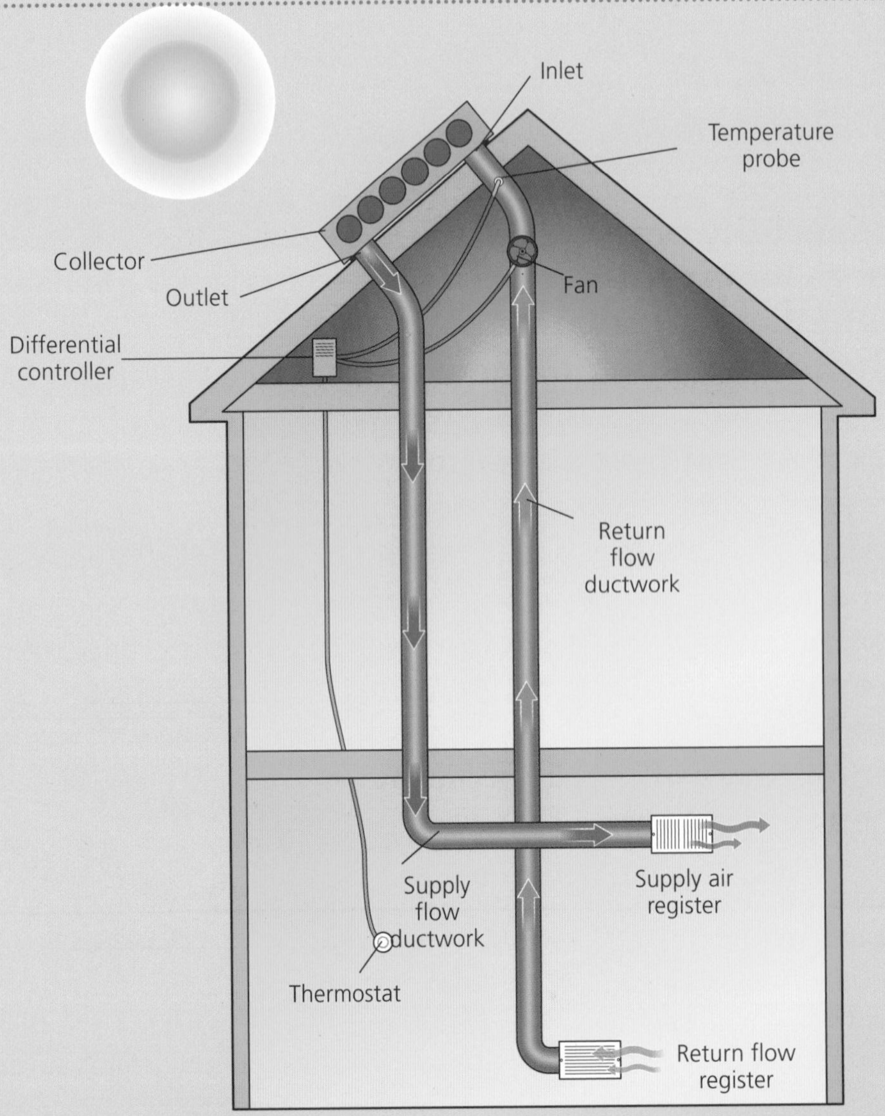

The **solar hot box** is a very simple system. Cold air from the house is drawn up into a network of ducts in the collector, where it is warmed by the sun then circulated inside to heat the house.

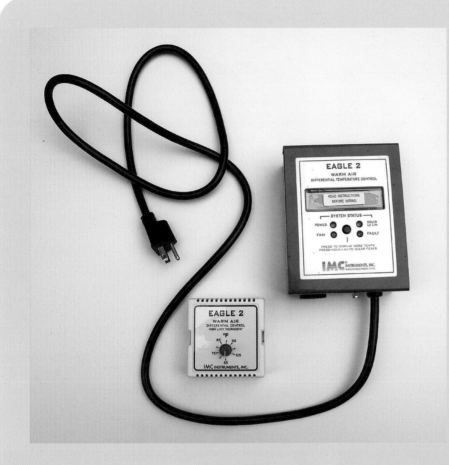

The temperature control equipment opens and closes the damper and causes the blower fan to turn on and off as needed.

TOOLS & MATERIALS

Jigsaw or circular saw with metal-cutting blade

Drywall saw

Straightedge

Aviation snips

Tape measure

Eye and ear protection

Carpenter's square

Drill/driver with bits

1/8" pop rivets (2)

Pop rivet gun

High-temperature silicone caulk

Caulk gun

Aluminum foil tape

1"-thick R7 rigid insulation

2 × 6 steel studs

Utility knife

Tempered glass

4" hole saw

Sheet-metal start collars

8" plenum box (2)

4" male and female duct connectors

1"-wide closed-cell foam gasket

4" aluminum HVAC duct

High-temperature black paint (matte)

Trim paint roller

Sheet-metal screws with rubber gaskets

Chalkline

Cardboard

Scissors

Reciprocating saw

Roof jack

Roofing cement

Flashing

Rubber gasket roofing nails

Shingles (if needed)

Unistrut

Unistrut connectors

Duct collars (2)

3/8" threaded rod

Spring-fed 8" backdraft dampers

8" blower fan

Temperature controls

Clamps

1" foam gasket tape

Work gloves

How to Build a Solar Hot Air Panel

Cut and bend the box frame pieces from 2 × 6 steel studs. Each steel stud piece will wrap two sides of the panel with a 90° corner bend. Mark the bend location on both steel studs. Cut a relief cut into the 6" side of the stud with aviation snips at this mark. Bend the stud to an L-shape and use a square to ensure that the corner forms a true 90° angle.

Drill ⅛"-dia. holes in the overlapping top and bottom flanges. Clamp the corners together before drilling and use a square to make sure the corner forms a 90° angle.

Fasten the corners of the metal box with two ⅛"-dia. sheet-metal pop rivets in the top and bottom. Leave one corner open to create access for the insulation panel insert.

Cut the foil-faced rigid foam insulation to match the interior dimensions of the box, using a drywall saw or a utility knife.

Apply high-temperature silicone to the bottom flanges of the box (inset). Fit the 1" foil-face rigid foam insulation into the back of the frame, then close up the box and secure the open corner. Cut 5"-wide strips of foam insulation to the length and width of the panel. Place a thick bead of silicone around the outside perimeter of the unit. Insert the strips into the silicone and tightly against the sides of the panel to hold the backing firmly in place. The foil should be facing into the box.

Seal the insulation edges. Place a bead of silicone around the inside corner where the insulation strips and backing panel meet, and then seal with foil tape. Flip the panel over. Place a bead of silicone on the intersection of the 2 × 6 stud flange and the back of the insulation and seal with foil tape. Conceal any exposed insulation edges with foil tape.

(continued)

How to Build a Solar Hot Air Panel (continued)

Create inlet and outlet holes in the walls with a hole saw or circle cutter. The number and location of the ductwork holes depends on where each panel fits into the overall array (presuming you are making and installing multiple panels). The first and last panels in the series each will have one end wall that is uncut, while intermediate panels will have duct holes on each end wall (inset).

Install a compartment separator in the first and last panels with a piece of foil insulation set on edge. Cut ductwork access holes in the separator. Then, cut out holes for the ductwork that will pass through the separator. Also cut a plenum opening in the separated compartment in the first and last unit.

Paint the entire box interior black using high-temperature paint and allow it to dry completely. A trim roller works well for this task.

Insert the ductwork. Beginning at the plenum over the inlet duct, guide 4" aluminum HVAC ductwork in a serpentine shape throughout the entire multi-panel installation, ending at the outlet duct. Join ends of adjoining duct sections with flexible duct connectors fashioned into a U shape and secured with metal screws and foil tape (inset). Paint each section of ductwork with black high-temperature paint once it is in place.

Paint the last section of ductwork and touch up around the interior of the box so all exposed surfaces are black.

Affix the glass top. First, double check that all openings in the panel are adequately sealed and insulated. Then, line the tops of the steel stud frame with foam closed-cell gasket tape. Carefully position the glass on top of the gasket tape, lined up ½" from the outside of the frame on all sides. Then, position foam closed-cell gasket tape around the perimeter of the top of the glass panel.

(continued)

How to Build a Solar Hot Air Panel (continued)

Attach the casing. Work with a local metal shop to bend metal flashing that will wrap your panel box. Attach around the perimeter of the panel with sheet-metal screws with rubber washer heads.
TIP: Be careful when working around the plenum ductwork. If you set the unit down on its backside, you will force the plenum up and break the seal around the opening.

Mark off the panel layout locations on the roof. Transfer the locations of the 8"-dia. inlet and outlet holes to the roof as well. The location of these holes should not interfere with the structural framing members of your roof (either rafters or trusses). Adjust the panel layout slightly to accommodate the best locations of the inlet and outlet, according to your roof's setup. Cut out the inlet and outlet holes with a reciprocating saw.

Use a roof jack or Cone-jack to form an 8"-dia. opening. Apply a heavy double bead of roofing cement along the top and sides of the roof jack. Nail the perimeter of the flange using rubber gasket nails. Cut and install shingles with roofing cement to fit over the flashing so they lie flat against the flange.

Attach Unistrut mounting U-channel bars to the roof for each panel. Use the chalklines on the roof to determine the position of the Unistrut, and attach to the roof trusses with Unistrut connectors.

Connect the panels to the Unistrut with ⅜" threaded rod attached at the top and bottom of the panel on the outside. Cut threaded rod to size, then attach to the Unistrut with Unistrut nuts. Attach the top clip to the top of the rod and the front face of the panel. Tighten the assembly to compress the panel down to the Unistrut for a tight hold. Seal the panel connections with 1" foam gasket tape around each end of the panels where they connect. Place a bead of silicone caulk on top of the gasket tape and then attach 3"-wide flashing over the two panels at the joint. Attach flashing to the panel with galvanized sheet metal screws with rubber gasket heads.

Hoist the panels into position. Carefully follow safety regulations and use scaffolding, ladders, ropes, and lots of helpers to hoist the panels onto the roof. Wear fall-arresting gear and take care not to allow the plenum ductwork to be damaged. **Connect the inlet** and outlet ducts on the panel(s) to the openings on the roof (inset). Position the panels so the inlet and outlet openings match perfectly, and attach with a duct collar and silicon caulk.

Hook up the interior ductwork, including dampers and a blower fan. The manner in which this is done will vary tremendously depending on your house structure and how you plan to integrate the supplementary heat. You will definitely want to work with a professional HVAC contractor (preferably one with experience with solar) for this part of the job.

wine racks for winemakers

40

Relaxing with a glass of wine—made from your own grapes, of course—is an important part of the self-sufficient lifestyle. Making wine from grapes and other fruit is too big a subject for this book, but in the following pages we'll show you some accessories for displaying and tasting your personal vintages.

Wine racks are the backbone of any wine cellar. If you are a skilled carpenter or woodworker, making your own wine racks is a fun exercise in designing and building. But if your ambition outpaces your experience, look into purchasing and installing a modular wine rack system. Sold over the Internet and at design centers, these systems allow you to design and install custom wine racks that fit your space, but at a fraction of the cost of hiring a professional carpenter to do the job. Most wine rack websites have planning software so you can create the exact design you want.

Starting with the dimensions of your cellar, including ceiling height, you can design a racking system with just the right bottle capacity and bonus features for you. The assembly of most kits of this type requires only few tools and little or no expertise. If you can read and follow instructions, you can build a modular wine rack. The model shown here (see Resources, page 348) relies on a system of ladders and latches for assembly.

Wine racks should be secured to the walls in the cellar, especially if you live in an earthquake-prone area. Attach the rack to wall studs or use appropriate hardware, such as hollow wall anchors or molly bolts, designed for the approximate weight of the loaded racks.

Any number of design and configuration options are possible with modular wine racks.

How to Build a Modular Wine Rack

Design your wine rack system and order the components. Typical components you can choose from include full-height racks with a separate cubby for each bottle; box or diamond-shaped racks; curved racks, quarter-round racks, corner racks, racks with tasting shelves, and more. Open the containers and inspect the parts when the kit arrives. Make sure there is no damage and that nothing is missing.

The package will include complete assembly instructions. Base your assembly on these. To build the arrangement shown here, we started with the full-height rack, identifying the ladder-shaped standards and orienting them with the bottom ends aligned. The standards should be set parallel on a flat surface.

Wood Selection Tips

Modular wine rack kits typically are sold in three or four wood species: cedar, redwood, red oak, and mahogany. The species you select will have a small effect on pricing, but the decision is primarily an aesthetic one. Western Red Cedar does not have an aromatic cedar scent, but is a clear, open-pored wood that produces a beautiful, mellow wood tone and does not require topcoating. Redwood is similar to cedar, but a bit denser and lighter in tone and with more limited availability. Red oak is harder and heavier and in most cases is stained and topcoated. Mahogany varies quite a bit, based on the country of origin (Malaysia and the Philippines, for example). It is a classic, open-pore wood with straight grain and good resistance to rot.

Attach spacer bars to the backs of the ladders at the prescribed rung locations, using finish nails or air-driven brad nails.

Lift the ladders and spacer bar assembly so it is upright. Insert intermediate ladders between the end ladders you have connected with the spacer bars, following the manufacturer's recommended spacing. Attach the intermediate ladders to the front spacer bars (as shown), according to the manufacturer's instructions.

Continue to build the structure by adding the next ladder, repeating steps 3 and 4.

Finish attaching the final front spacer bars, and then move the unit into the desired place against the wall. Attach the assembly to the wall using 2½" screws driven through the back spacer bars. Make sure the assembly is level first, and drive the screws at wall stud locations (or use masonry anchors if walls are made of concrete or block).

Attach the last modular unit according to the installation instructions. Also install any trim pieces to conceal gaps between units and between the end unit and the wall. Most wood modular rack systems are either prefinished or designed to remain unfinished. Begin loading your wine collection into the racks.

Building a Countertop Wine Rack

Large wine racks have their places; so do smaller ones. You may want to showcase certain bottles or simply have them close at hand during a party or wine tasting. The rack shown in this project holds eight bottles and would make a nice gift for a new collector or even a noncollector who just likes to have a few bottles of wine at the ready.

We used clear birch for the rack, but any finish-grade, surface-planed hardwood will work just as well. And although we used a finish with a stain, you may choose to stain the wood first or even paint the rack. This is a matter of pure preference.

When building this rack, use regular carpenter's glue rather than the high-strength type. The regular type has a longer open time, which you'll probably need to get the entire rack assembled. If glue oozes out around the holes, wipe it off with a damp paper towel right away.

A countertop wine rack is an excellent way to showcase special bottles or to keep favorites within easy reach.

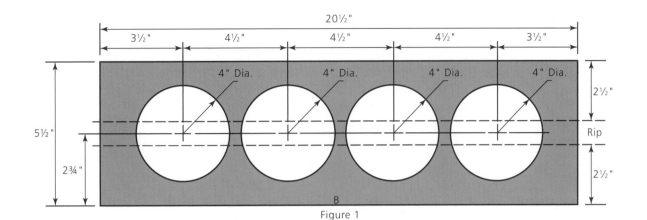

Figure 1

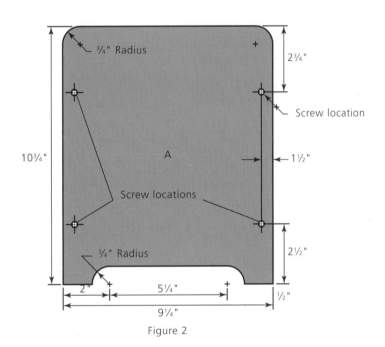

Figure 2

TOOLS & MATERIALS

Circular saw
Jigsaw or coping saw
Drill
Rubber mallet
(3) Bar clamps
Transfer paper

#8 × 1¼" wood screws
Wood glue
Medium- and fine-grit sanding sponge
Polyurethane finish
Paintbrush
2 lineal' hardwood 1 × 10

4 lineal' hardwood 1 × 6
³/₈" dowel plugs
Router with ¼" roundover bit
Eye and ear protection
Work gloves

CUTTING LIST

KEY	No.	PART	DIMENSION	MATERIAL
A	2	Sides	¾ × 9¼ × 10¾"	Hardwood
B	2	Shelf front/back	¾ × 5½ × 20½"	Hardwood

How to Build a Countertop Wine Rack

Prepare stock as necessary for milling (we are using premilled 1 × 10 and 1 × 6 birch here—actual size is ¾" thick and 9¼" wide and 5½" wide). Cut the 1 × 10 board into two pieces, each 10¾" long. Cut the 1 × 6 into two pieces, each 20½" long.

Using a photocopier, enlarge the patterns on page 269. Transfer the rack pattern onto your 1 × 6 wood stock, using transfer paper and a stylus or hard pencil. Drill a ¼" hole near the edge of each circle, then slip the jigsaw blade into the hole and cut along marked lines to cut out the circles.

Rip-cut each rack piece along the marked line from the pattern, using a circular saw and straightedge guide or a table saw.

Transfer the side pattern onto each piece of 1 × 10. Mark the screw locations (see diagram on page 269) too. Cut along marked lines, using a jigsaw.

Use a router and a ¼" roundover bit to shape the edges of the panels (except the bottoms of the feet) and the edges of the racks (but not the ends). Sand the faces and edges of all the pieces.

Place end panels face-up on a workbench. Drill ⅜" countersunk pilot holes at marked screw locations. Spread glue on the ends of the rack pieces, assemble the pieces, and clamp the assembly together, using bar or pipe clamps.

Drive screws through the pilot holes in the panels and into the rack pieces. Spread glue on the ends of ⅜" dowel plugs or buttons and insert one into each screw hole to conceal the screw heads. Trim and sand the plugs flush after the glue dries.

When the glue is dry, sand the entire unit with a fine-grit sanding sponge and apply two coats of a finish of your choice.

Building a Tasting Table

Some wine cellars are strictly functional: that is, they are temperature- and humidity-controlled repositories for your cherished bottles of wine. But a wine cellar, especially a home wine cellar, can be much more than that. If you aren't thinking about including a table and perhaps seating in your wine cellar, you're missing an opportunity to create a room of high romance and appeal.

Tasting tables are one of the more popular members of the wine furniture family. Often, they are nothing more than an old barrel or cask set on end. All that is required is a small, sturdy surface where you can uncork a bottle and sample a glass or two. Some tasting tables have matching seating (usually bar stool height), while others are built with the presumption that busy wine tasters prefer a surface at a height that lends itself to imbibing while standing upright.

If you do a little searching around with wine furniture purveyors, you'll find that tasting tables often include wine racks in the table base. The tabletops are typically round in the French bistro fashion, and the prices range from a couple hundred dollars for flimsier models to a thousand or more for models with beefy hardwood butcher-block tops and very burly construction. The design you see here can be constructed with less than $30 worth of materials, but with a nice rich finish that will exhibit the Old World charm we've come to expect even in a contemporary wine cellar.

The tabletop is fashioned from ordinary SPF (spruce, pine, or fir) 2 × 4 stock that has been ripped down to 2½ inches wide to remove the bullnosed edges and give the table a slightly more streamlined appearance. The 2 × 4 stock is face glued to create a 24 × 24-inch square, from which the 24-inch-diameter round top is cut. Then, the top is banded with ⅛-inch strap iron. This is mostly for visual purposes, but the banding will help keep the top from warping as it expands and contracts.

If you'd prefer and you have a couple hundred dollars to spend, you can buy a preformed round butcher block. If you go this route, you might as well invest in the blocks that are a full 3 inches thick. And look for a maple or beech block that is formed with the end grain of the individual blocks pointing upward.

The table base is made from ordinary 1 × 4 pine and 21-inch-diameter MDF discs.

This lovely two-level tasting table adds a gracious touch to a wine cellar and provides a place to serve and enjoy a glass of your favorite vintage.

TOOLS & MATERIALS

Table saw
Jigsaw or bandsaw
Compass or trammel points
Power sander
Drill
Hacksaw
Nut driver
Paintbrush
(20) black 1½" hex head
    lag screws
8d finish nails or 2"
    pneumatic nails
Wood glue
Shellac
Dark wood stain
Enamel paint
⅛ × 1½ × 75½" strap iron
(4) 2 × 4" × 8' SPF
(2) 1 × 4" × 8' pine
(½ sheet) ¾" MDF
3" wood screws
2½" wood screws
Flat washers
1¼" brass wood screws
Clamps
Eye and ear protection
Work gloves

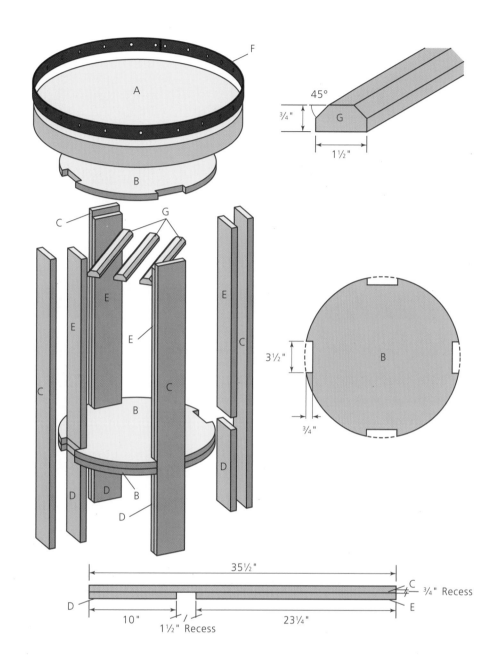

CUTTING LIST

KEY	No.	PART	DIMENSION	MATERIAL
A	1	Tabletop	2½ × 24" dia.	1 × 4 Poplar
B	3	Shelf disc	¾ × 21" dia.	MDF
C	4	Leg	¾ × 3½ × 35½"	Pine
D	4	Filler-short	¾ × 3½ × 10"	Pine
E	4	Filler-long	¾ × 3½ × 23¼"	Pine
F	1	Metal strap	⅛ × 1½ × 75½"	Strap metal (cold rolled)
G	3	Stemware racks	¾ × 1½ × 12"	Oak (hardwood)

How to Build a Tasting Table

Select four 8' 2 × 4s that are clear and straight, and then rip-cut them to 2½" wide by trimming ½" off each edge on a table saw to remove the bullnose profiles. Cut them to length (24") and lay the 16 workpieces face-to-face on a flat surface, forming a 24 × 24" square (A). Then, apply liberal amounts of wood glue to both surfaces and clamp the workpieces together firmly with several pipe or bar clamps. Let the glue-up cure overnight.

Use a compass or trammel points to lay out a 24"-dia. circle onto the tabletop glue-up. To make your own compass, tie a pencil to one end of a piece of twine, and then tie the other end of the twine to a nail so the nail and pencil point are exactly 12" apart when the twine is pulled taut. Drive the nail in the center of the glue-up. Then, pull the twine taut and trace a circular cutting line onto the glue-up.

Cut out the round shape using a bandsaw, if you have one, or a jigsaw with a wide, stiff wood-cutting blade. Cut just outside the cutting line so you can sand the edge up to the line on a stationary sander or with a belt sander.

Lightly resurface the tabletop with a belt sander and 150-grit sanding belt. This will get rid of dried glue squeeze-out and create a smoother, more even surface. Also finish-sand the edges.

Apply a coat of thinned shellac to the completely sanded tabletop as a sanding sealer. Let the sealer dry and then stain the tabletop with a dark wood stain, such as dark walnut. Once you have achieved the color tone you like, apply a penetrating topcoat if desired (such as wipe-on tung oil).

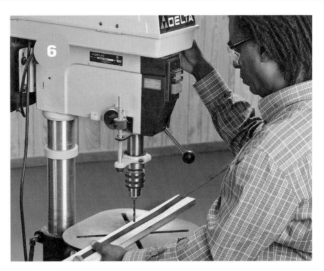

Purchase a piece of ⅛"-thick × 1½"-wide strap iron that's at least 76" long and cut it to length (75½", but measure the circumference of the top first to confirm). Also purchase square-head screws, preferably decorative black lag-style, and then drill a screw hole every 4" in the strap metal. Use a drill press when possible and lubricate the drilling area with cutting oil. Paint the strap with enamel paint.

Center the strap on the edge of the tabletop and mark a pilot hole for the first screw. Start in the middle of the strap (end to end). Drive a hex-head screw to fasten the strap. Then, add the next screw in line, and then the next on the opposite side. Alternate back and forth, bending the strap as you go so it is flush against the wood.

Cut the 1 × 4 stock for the legs, according to the cutting list. The legs are made by face-gluing the fillers (D, E) onto the inside faces of the full-length legs (C), creating a ¾" recess at the top and a 1½" recess starting 10" up from the bottom. Lay out the legs in a row so you know they're even, and attach the filler strips with glue and finish nails or pneumatic nails.

(continued) 275

How to Build a Tasting Table (continued)

Cut three 21"-dia. round shelves (B) from ¾"-thick MDF (medium-density fiberboard). Use the same compass technique shown in step 2 and the cutting technique shown in step 3. Once all three circles are cut, gang them together and sand them so the edges are round, smooth, and uniform.

Keeping the three circles ganged together, bisect the top circle to create four equal quadrants. Mark a centerpoint on the end of a short piece of 1 × 4 scrap. Then, position the scrap so the centerpoint aligns with each quadrant line and the back face of the scrap is flush with the edge of the disc assembly. Trace the scrap piece to create four notches for the 1 × 4 legs.

Cut out the notches with a jigsaw or bandsaw and sand them so they're smooth and square. Un-gang the three-circle assembly, and fill any nail or screw holes you may have created by ganging the parts.

Drill counterbored pilot holes for 3" wood screws through the leg notch areas and into the shelf discs. Attach the legs to the shelves with glue and screws. Test to make sure the base is level and square before filling the counterbores with wood plugs or wood filler.

Finish-sand the legs and the circular shelves and apply the finish. Here, the same finish that's applied to the tabletop is applied to both the legs and the shelves (a sealer coat of thinned shellac followed by dark walnut stain and then penetrating oil). Because the stain is dark, the MDF shelves accept it well and blend with the legs and tabletop. But you may prefer to paint the shelves gloss black or even a metallic tone.

Set the tabletop good-side-down on a flat surface and then arrange the top of the base on the underside of the tabletop. The top should be centered, with an equal overhang (around 1½") all around. Drill several ¼"-dia. access holes through the top shelf in the base, but not into the tabletop. Then, drill pilot holes into the tabletop, centered on the access holes. Slip metal washers onto 2½" wood screws and drive one at each hole (using this washer and guide hole system allows for some wood movement).

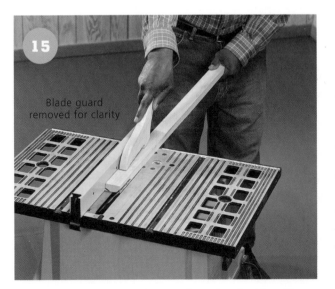

Blade guard removed for clarity

Make a stemware rack to hang on the underside of the tabletop. Cut a 45° bevel that starts halfway up each edge of a strip of 1 × 2 hardwood. Cut the strip into three 12" lengths (G).

Arrange the three strips in parallel configuration on the underside of the top shelf. Use a wine glass as a reference for spacing the strips to create the rack. Keep the ends flush, and then attach the strips with counterbored 1¼" brass wood screws.

outbuildings
& fences

Anyone dedicated to the self-sufficient lifestyle will soon discover that they need additional storage space for equipment, sheltered growing spaces for plants to increase the size of the harvest, a shed or two, maybe a small pole barn, fences to keep livestock in and deer out—the list goes on and on. You can spend a lot of money building these structures, but you don't have to.

In this section we'll first walk you through some inexpensive plans that can help you grow more produce by extending the growing season and protecting delicate plants from harsh weather conditions. Both our greenhouse and hoophouse are designed to fit into small yards, but they're really just downsized versions of structures used on commercial farms, so they can easily be scaled up if space allows. You can also cover them with heavier duty materials such as doublewall polycarbonate sheets or other commercial-grade coverings.

As you get more involved in self-sufficiency, you'll start accumulating more tools and equipment, and tools and equipment require storage. If you have the time and funds, consider building a pole barn, which will give you plenty of room for storage and workspace, and probably enough room for expansion in the years to come (who knows, you may need a tractor someday). However, if you just need space for garden equipment, a metal kit shed, which has precut parts and is easy to assemble, can be the perfect solution. We show plans for both alternatives in this section.

Fences are another common exterior building project, and they quickly become a necessity when you have animals or need to protect your produce from ravenous local deer. It's important that they be functional and do the job they're intended for, but there's a wide range of options and looks, from sturdy chain-link fencing to old-fashioned split rail fencing.

greenhouse

41

Utilizing a greenhouse is a great way to extend and diversify your garden, enabling you to grow more food for a longer period of time— and perhaps even grow foods that wouldn't otherwise survive in your climate. A greenhouse can be a decorative and functional building that adds beauty to your property, or a quick and easy temporary structure that serves a purpose and then disappears. The wood-framed greenhouse in this project is somewhere between these two types. The sturdy wood construction will hold up for many seasons. The plastic sheeting covering will last one to five seasons, depending on the material you choose, and is easy to replace when it starts to degrade.

The 5-foot-high kneewalls in this design provide ample space for installing and working on a conventional-height potting table. For a door, this plan simply employs a sheet of weighted plastic that can be tied out of the way for entry and exit. If you plan to go in and out frequently, you can purchase a prefabricated greenhouse door from a garden center or greenhouse materials provider. To allow for ventilation in hot weather, we built a wood-frame vent cover that fits over one rafter bay and can be propped open easily.

A wood-frame greenhouse with sheet-plastic cover is an inexpensive, semi-permanent gardening structure that can be used as a potting area as well as a protective greenhouse.

Where to Site Your Greenhouse

When the first orangeries (early greenhouses) were built, heat was thought to be the most important element for successfully growing plants indoors. Most orangeries had solid roofs and walls with large windows. Once designers realized that light was more important than heat for plant growth, they began to build greenhouses from glass.

All plants need at least 6 (and preferably 12) hours of light a day year-round, so when choosing a site for a greenhouse, you need to consider a number of variables. Be sure that it is clear of shadows cast by trees, hedges, fences, your house, and other buildings. Don't forget that the shade cast by obstacles changes throughout the year. Take note of the sun's position at various times of the year. A site that receives full sun in the spring and summer can be shaded by nearby trees when the sun is low in winter. Winter shadows are longer than those cast by the high summer sun, and during winter, sunlight is particularly important for keeping the greenhouse warm. If you are not familiar with the year-round sunlight patterns on your property, you may have to do a little geometry to figure out where shadows will fall. Your latitude will also have a bearing on the amount of sunlight available; greenhouses at northern latitudes receive fewer hours of winter sunlight than those located farther south. You may have to supplement natural light with interior lighting.

To gain the most sun exposure, the greenhouse should be oriented so that its ridge runs east to west (see illustration), with the long sides facing north and south. A slightly southwest or southeast exposure is also acceptable, but avoid a northern exposure if you're planning an attached greenhouse; only shade-lovers will grow there.

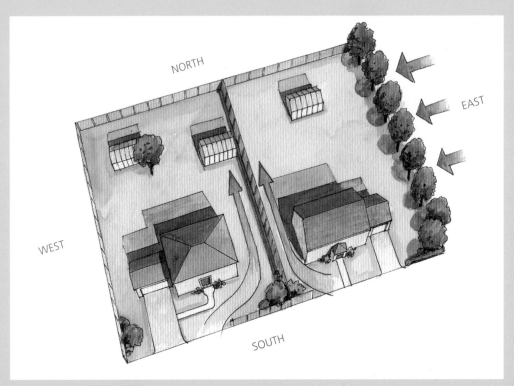

The ideal greenhouse location is well away from trees but protected from prevailing winds, usually by another structure, a fence or a wall.

For maximum heat gain, orient your greenhouse so the roof or wall with the most surface area is as close to perpendicular to the sunrays as it can be.

Building a Greenhouse

CUTTING LIST

Key	No.	Part	Dimension	Material
A	2	Base ends	3½ × 3½ × 96"	4 × 4 landscape timber
B	2	Base sides	3½ × 3½ × 113"	4 × 4 landscape timber
C	2	Sole plates end	1½ × 3½ × 89"	2 × 4 pressure treated
D	2	Sole plates side	1½ × 3½ × 120"	2 × 4 pressure treated
E	12	Wall studs side	1½ × 3½ × 57"	2 × 4
F	1	Ridge support	1½ × 3½ × 91"	2 × 4
G	2	Back studs	1½ × 3½ × 76" *	2 × 4
H	2	Door frame sides	1½ × 3½ × 81" *	2 × 4
I	1	Cripple stud	1½ × 3½ × 16"	2 × 4
J	1	Door header	1½ × 3½ × 32"	2 × 4
K	2	Kneewall caps	1½ × 3½ × 120"	2 × 4
L	1	Ridge pole	1½ × 3½ × 120"	2 × 4
M	12	Rafters	1½ × 3½ × 60" *	2 × 4

*Approximate dimension; take actual length and angle measurements on structure before cutting.

TOOLS & MATERIALS

(1) 20 × 50' roll 4- or 6-mil polyethylene sheeting or greenhouse fabric, tack strips

(12) 24"-long pieces of No. 3 rebar

(4) 16' pressure-treated 2 × 4

(2) exterior-rated butt hinges

(1) screw-eye latch

(8) 8" timber screws

Reciprocating saw

Level

Carpenter's square

Drill with a nut driver bit

Eye and ear protection

Metal cutoff saw

Maul or sledgehammer

Speed square

3" deck screws

Exterior panel adhesive

Caulk gun

Hammer

Jigsaw or handsaw

Miter saw

Circular saw

Utility knife

Pneumatic nailer

Scissors

Scrap 2 × 4

Pencil

Ladder

Screen retaining strips

Work gloves

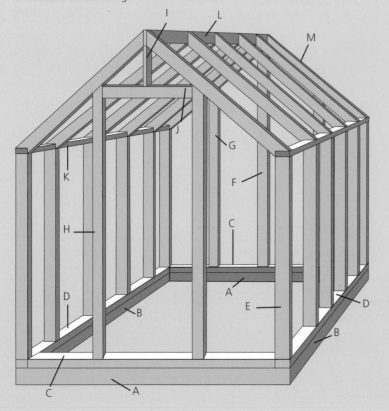

How to Build a Greenhouse

Prepare the installation area so it is flat and well drained; then cut the base timbers (4 × 4 landscape timbers) to length. Arrange the timbers so they are flat and level and create a rectangle with square corners. Drive a pair of 8" timber screws at each corner, using a drill/driver with a nut-driver bit.

Cut 12 pieces of No. 3 rebar (find it in the concrete supplies section of any building center) to 24" in length to use as spikes for securing the timbers to the ground. A metal cutoff saw or a reciprocating saw with a metal cutting blade can be used to make the cuts. Drill a ⅜" guide hole through each timber near each end and in the middle. Drive a rebar spike at each hole with a maul or sledgehammer until the top is flush with the wood.

Cut the plates and studs for the two side walls (called knee-walls). Arrange the parts on a flat surface and assemble the walls by driving three 3" deck screws through the cap and base plates and into the ends of the studs. Make both kneewalls.

(continued)

How to Build a Greenhouse (continued)

Set the base plate of each kneewall on the timber base and attach the walls by driving 3" deck screws down through the base plates and into the timbers. For extra holding power, you can apply exterior panel adhesive to the undersides of the plates, but only if you don't plan to relocate the structure later.

Cut the ridge support post to length and attach it to the center of one end base plate, forming a T. Cut another post the same length for the front (this will be a temporary post) and attach it to a plate. Fasten both plates to front and back end timbers.

Set the ridge pole on top of the posts and check that it is level. Also check that the posts are level and plumb. Attach a 2 × 4 brace to the outer studs of the kneewalls and to the posts to hold them in square relationship. Double-check the pole and posts with the level.

Cut a 2 × 4 to about 66" to use as a rafter template. Hold the 2 × 4 against the end of the ridge pole and the top outside corner of a kneewall. Trace along the face of the ridge and the top plate of the wall to mark cutting lines. Cut the rafter and use it as a template for the other rafters on that side of the roof. Create a separate template for the other side of the roof.

Mark cutting lines for the rafters using the templates, and cut them all. You'll need to use a jigsaw or handsaw to make the bird's mouth cuts on the rafter ends that rest on the kneewall.

Attach the rafters to the ridge pole and the kneewalls with deck screws driven through pilot holes. Try to make the rafters align with the kneewall studs.

(continued)

How to Build a Greenhouse (continued)

Mark the positions for the remaining end wall studs on the base plate. At each location, hold a 2 × 4 on end on the base plate and make it level and plumb. Trace a cutting line at the top of the 2 × 4 where it meets the rafter. Cut the studs and install them by driving screws toenail-style.

Measure up 78" (or less if you want a shorter door) from the sole plate in the door opening and mark a cutting line on the temporary ridge post. Make a square cut along the line with a circular saw or cordless trim saw (inset). Then cut the door header to fit between the vertical door frame members. Screw the header to the cut end of the ridge post and drive screws through the frame members and into the header.

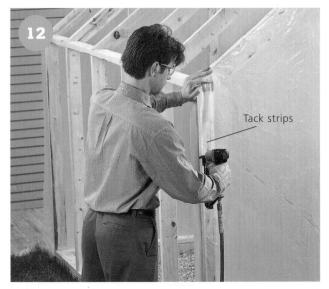

Tack strips

Begin covering the greenhouse with your choice of cover material. (We used 6-mil polyethylene sheeting.) Start at the ends. Cut the sheeting to size and then fasten it by attaching screen retainer strips to wood surfaces at the edges of the area being covered. Tack the sheeting first at the top, then at the sides and finally at the bottom. After the strips are installed (use wire brads), trim the sheeting along the edges of the strips with a utility knife.

Attach the sheeting to the outside edge of the base plate on one side. Roll sheeting over the roof and down the other side. Draw it taut and cut it slightly overlong with scissors. Attach retainer strips to the other base plate and then to the outside edges of the corner studs.

Make and hang a door. We simply cut a piece of sheet plastic a little bigger than the opening (32") and hung it with retainer strips from the header. Attach a piece of 2 × 4 to the bottom of the door for weight.

OPTION: Make a vent window. First, cut a hole in the roof in one rafter bay and tack the cut edges of the plastic to the faces (not the edges) of the rafters, ridge pole, and wall cap. Then build a frame from 1 × 2 stock that will span from the ridge to the top of the kneewall and extend a couple of inches past the rafters at the side of the opening. Clad the frame with plastic sheeting and attach it to the ridge pole with butt hinges. Install a screw-eye latch to secure it at the bottom. Make and attach props if you wish.

hoophouse

42

The hoophouse is a popular garden structure for two main reasons: It is cheap to build and easy to build. In many agricultural areas you will see hoophouses snaking across vast fields of seedlings, protecting the delicate plants at their most vulnerable stages. Because they are portable and easy to disassemble, they can be removed when the plants are established and less vulnerable.

While hoophouses are not intended as inexpensive substitutes for real greenhouses, they do serve an important agricultural purpose. And building your own is a fun project that the whole family can enjoy.

The hoophouse shown here is essentially a Quonset-style frame of bent ¾" PVC tubing draped with sheet plastic. Each semicircular frame is actually made from two 10-foot lengths of tubing that fit into a plastic fitting at the apex of the curve. PVC tubes tend to stay together simply by friction-fitting into the fittings, so you don't normally need to solvent glue the connections (this is important to the easy-to-disassemble and store feature). If you experience problems with the frame connections separating, try cutting 4- to 6-inch-long pieces of ½-inch (outside diameter) PVC tubing and inserting them into the tubes and fittings like splines. This will stiffen the connections.

A hoophouse is a temporary agricultural structure designed to be low-cost and portable. Also called Quonset houses and tunnel houses, hoophouses provide shelter and shade (depending on the film you use) and protection from wind and the elements. They will boost heat during the day but are less efficient than paneled greenhouses for extending the growing season.

Instead of plastic tubing, you may use green tree branches to create the ribs for your hoophouse. Look for branches no more than 1" in diameter and 10' long or so. If you can't find branches that are long enough, lash them together at the ridge (inset).

Building & Siting a Hoophouse

The fact that a hoophouse is a temporary structure doesn't give you license to skimp on the construction. When you consider how light the parts are and how many properties sheet plastic shares with boat sails, the importance of securely anchoring your hoophouse becomes obvious. Use long stakes (at least 24 inches) to anchor the tubular frames, and make sure you have plenty of excess sheeting at the sides of the hoophouse so the cover can be held down with ballast. Creating pockets at the ends of the sheeting and inserting scrap lumber is the ballasting technique shown here, but it is also common (especially when building in a field) to weigh down the sheeting by burying the ends in dirt. Only attach the sheeting at the ends of the tubular frame, and where possible, orient the structure so the prevailing winds won't blow through the tunnel.

Building a Hoophouse

- Space frame hoops about 3' apart.

- Leave ridge members a fraction of an inch (not more than ¼") shorter than the span, which will cause the structure to be slightly shorter on top than at the base. This helps stabilize the structure.

- Orient the structure so the wall faces into the prevailing wind rather than the end openings.

- If you are using long-lasting greenhouse fabric for the cover, protect the investment by spray-painting the frame hoops with primer so there is no plastic-to-plastic contact.

- Because hoophouses are temporary structures that are designed to be disassembled or moved regularly, you do not need to include a base.

- The ¾" PVC pipes used to make the hoop frames are sold in 10' lengths. Two pipes fitted into a tee or cross fitting at the top will result in legs that are 10' apart at the base and a ridge that is roughly 7' tall.

- Clip the hoophouse covers to the end frames. Clips fastened at the intermediate hoops will either fly off or tear the plastic cover in windy conditions.

Row tunnels are often used in vegetable gardens to protect sensitive plants in the spring and fall. Plastic or fabric sheeting is draped over a short wire or plastic framework to protect plants at night. During the heat of the day, the sheeting can be drawn back to allow plants direct sunlight.

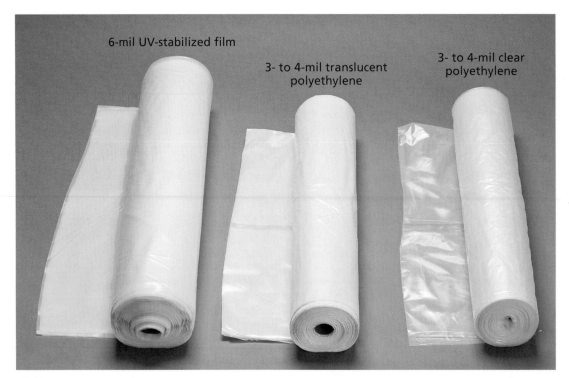

Sheet plastic is an inexpensive material for creating a greenhouse. Obviously, it is less durable than polycarbonate, fiberglass, or glass panels. But UV-stabilized films at least 6-mil thick can be rated to withstand four years or more of exposure. Inexpensive polyethylene sheeting (the kind you find at hardware stores) will hold up for a year or two, but it becomes brittle when exposed to sunlight. Some greenhouse builders prefer to use clear plastic sheeting to maximize the sunlight penetration, but the cloudiness of translucent poly makes it effective for diffusing light and preventing overheating. For the highest quality film coverings, look for film rated for greenhouse and agricultural use.

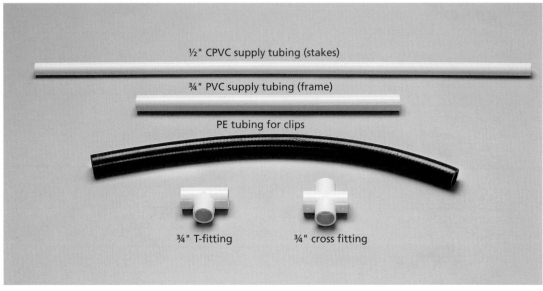

Plastic tubing and fittings used to build this hoophouse include: Light duty ¾" PVC tubing for the frame (do not use CPVC—it is too rigid and won't bend properly); ½" CPVC supply tubing for the frame stakes (rigidity is good here); Polyethylene (PE) tubing for the cover clips; T-fittings and cross fittings to join the frame members.

Building a Hoophouse

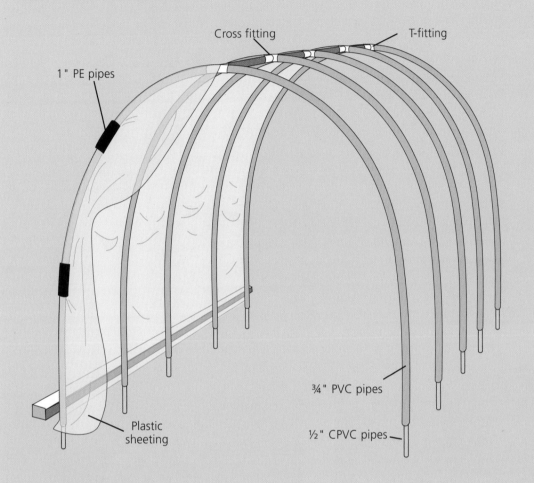

Cross fitting

T-fitting

1" PE pipes

¾" PVC pipes

½" CPVC pipes

Plastic
sheeting

TOOLS & MATERIALS
(for 10' wide × 12' long project seen here)

(12) ¾" × 10' PVC pipes

(3) ½" × 10' CPVC pipes

(1) 1" × 10' PE pipe (black)

(3) ¾" PVC cross fittings

(2) ¾" PVC T-fittings

20 × 16' clear or translucent
   plastic sheeting

(4) 12' pressure-treated 2 × 4

Stakes

Eye and ear protection

Mason's string

Tape measure

Circular saw

Painters' tape

Mallet

Maul

Stapler

2½" deck screws

Drill

Work gloves

How to Build a Hoophouse

Lay out the installation area using stakes and mason's string. Strive for square corners, but keep in mind that these are relatively forgiving structures, so you can miss by a little bit and probably won't be able to notice.

Cut a 30"-long stake from ½" CPVC supply tubing for each leg of each hoop frame. Measure out from the corners of the layout and attach a piece of high-visibility tape on the string at 3' intervals; then drive a stake at each location. When the stake is fully driven, 10" should be above ground and 20" below.

Join the two legs for each frame hoop with a fitting. Use a T-fitting for the end hoop frames and a cross fitting for the intermediate hoop frames. No priming or solvent gluing is necessary. (The friction-fit should be sufficient, but it helps if you tap on the end of the fitting with a mallet to seat it.)

Slip the open end of one hoop-frame leg over a corner stake so the pipe is flush against the ground. Then bend the pipes so you can fit the other leg end over the stake at the opposite corner. If you experience problems with the pipes pulling out of the top fitting, simply tape the joints temporarily until the structure frame is completed.

Continue adding hoop frames until you reach the other end of the structure. Wait until all the hoop frames are in place before you begin installing the ridge poles. Make sure the cross fittings on the intermediate hoop frames are aligned correctly to accept the ridge poles.

Add the ridge-pole sections between the hoop frames. Pound on the end of each new section as you install it to seat it fully into the fitting. Install all of the poles.

(continued)

How to Build a Hoophouse (continued)

Cut four pieces of pressure-treated 2 × 4 to the length of the hoophouse (12' as shown). Cut the roof cover material to size. (We used 6-mil polyethylene sheeting.) It should be several inches longer than is necessary in each direction. Tack the cover material at one end of the 2 × 4 and then continue tacking it as you work your way toward the end. Make sure the material stays taut and crease-free as you go.

Lay a second 2 × 4 the same length as the first over the tacked plastic so the ends and edges of the 2 × 4s are flush. Drive a 2½" deck screw through the top 2 × 4 and into the lower one every 24" or so, sandwiching the cover material between the boards. Lay the assembly next to one edge of the hoophouse and pull the free end of the material over the tops of the frames.

On the other side of the structure, extend the cover material all the way down so it is taut and then position another 2 × 4 underneath the fabric where it meets the ground. Staple the plastic and then sandwich it with a final 2 × 4.

Make clips to secure the roof cover material from a 12"-long section of hose or soft tubing. Here, 1"-dia., thin-walled PE supply tubing is slit longitudinally and then slipped over the material to clip it to the end frames. Use at least six clips per end. Do not clip at the intermediate hoop frames.

OPTION: Make doors by clipping a piece of cover material to each end. (It's best to do this before attaching the main cover.) Then cut a slit down the center of the end material. You can tie or tape the door material to the sides when you want it open and weigh down the pieces with a board or brick to keep the door shut. This solution is low-tech but effective.

metal kit shed

43

You don't have to have construction expertise to build a good-looking and useful addition to your yard. If you need an outbuilding but don't have the time, expertise, or inclination to build one from scratch, a kit shed is the answer. Today's kit sheds are available in a wide range of materials, sizes, and styles—from snap-together plastic lockers to Norwegian pine cabins with divided-light windows and loads of architectural details.

Equally diverse is the range of quality and prices for shed kits. One thing to keep in mind when choosing a shed is that much of what you're paying for is the materials and the ease of installation. Better kits are made with quality, long-lasting materials, and many come largely preassembled. A shed from a kit is a known quantity—the look, the storage space, and the complexity of it are apparent beforehand.

The metal shed described here is typical. It measures 8 × 9 and comes complete with every piece you'll need for the main building precut and predrilled. All you need is a ladder and a few hand tools for assembly. The pieces are lightweight and maneuverable, but it helps to have at least two people for fitting everything together.

As with most kits, this shed does not include a foundation as part of the standard package. It can be built on top of a patio surface or out in the yard, with or without an optional floor. To help keep it level and to reduce moisture from ground contact, it's a good idea to build it over a bed of compacted gravel. A 4-inch-deep bed that extends about 6 inches beyond the building footprint makes for a stable foundation and helps keep the interior dry throughout the seasons.

Before you purchase a shed kit, check with your local building department to learn about restrictions that affect your project. It's recommended—and often required—that lightweight metal sheds be anchored to the ground. Shed manufacturers offer different anchoring systems, including cables for tethering the shed into soil, and concrete anchors for tying into a concrete slab.

A shed like this one is an incredibly handy storage feature for just about any homestead. It can be used to store game feed, yard tools and supplies, or even firewood.

Building a Kit Shed

Wooden shed kits generally follow the same, simple-to-build construction style. Using the structural gussets and framing connectors, you avoid tricky rafter cuts and roof assembly. Many hardware kits come with lumber cutting lists so you can build the shed to the desired size without using plans.

Metal doesn't mean ugly when it comes to kit sheds. As this unit clearly shows, these homestead additions add both abundant space and understated style that doesn't detract from the look of the yard.

Shed Features to Consider

Here are some of the key elements to check out before purchasing a kit shed.

Materials

Shed kits are made of wood, metal, vinyl, various plastic compounds, or any combination thereof. Consider aesthetics, of course, but also durability and appropriateness for your climate. For example, check the snow load rating on the roof if you live in a snowy climate, or inquire about the material's UV resistance if your shed will receive heavy sun exposure. The finish on metal sheds is important for durability. Protective finishes include paint, powder-coating, and vinyl. For wood sheds, consider all of the materials, from the framing to the siding, roofing, and trimwork.

Extra Features

Do you want a shed with windows or a skylight? Some kits come with these features, while others offer them as optional add-ons. For a shed workshop, office, or other workspace where you'll be spending a lot of time, consider the livability and practicality of the interior space, and shop accordingly for special features.

What's Included?

Many kits do not include foundations or floors, and floors are commonly available as extras. Other elements you're not likely to find are roof coverings (often the plywood roof sheathing is included but not the building paper, drip edge, or shingles); paint or finish (some sheds come prefinished though). Most shed kits include hardware (nails, screws) for assembling the building, but always check this to make sure. Easy-assembly models may have wall siding and roof shingles already installed onto panels.

Prepare the building site by leveling and grading as needed, and then excavating and adding a 4"-thick layer of compactible gravel. If desired, apply landscape fabric under the gravel to inhibit weed growth. Compact the gravel with a tamper and use a level and a long, straight 2 × 4 to make sure the area is flat and level.

Begin by assembling the floor according to the manufacturer's directions—these will vary quite a bit among models, even within the same manufacturer. Be sure that the floor system parts are arranged so the door is located where you wish it to be. Do not fasten the pieces at this stage. **NOTE:** Always wear work gloves when handling shed parts—the metal edges can be very sharp.

Once you've laid out the floor system parts, check to make sure they're square before you begin fastening them. Measuring the diagonals to see if they're the same is a quick and easy way to check for square.

Fasten the floor system parts together with kit connectors once you've established that the floor is square. Anchor the floor to the site if your kit suggests. However, some kits are designed to be anchored after full assembly is completed. Follow the manufacturer's recommendations.

(continued)

Building a Kit Shed (continued)

Begin installing the wall panels according to the instructions. Most panels are predrilled for fasteners, so the main trick is to make sure the fastener holes align between panels and with the floor.

Tack together mating corner panels on at least two adjacent corners. If your frame stiffeners require assembly, have them ready to go before you form the corners. With a helper, attach the frame stiffener rails to the corner panels.

Install the remaining fasteners at the shed corners once you've established that the corners all are square.

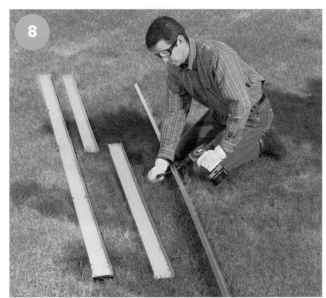

Lay out the parts for assembling the roof beams and the upper side frames and confirm that they fit together properly. Then, join the assemblies with the fasteners provided.

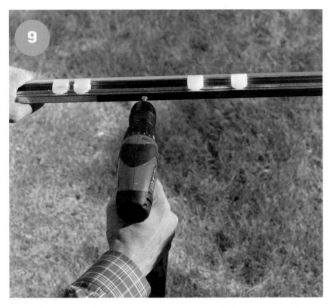

Attach the moving and nonmoving parts for the upper door track to the side frames if your shed has sliding doors.

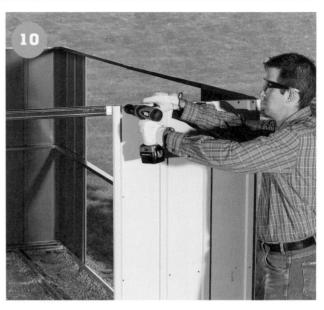

Fasten the shed panels to the top frames, making sure that any fastener holes are aligned and that crimped tabs are snapped together correctly.

Fill in the wall panels between the completed corners, attaching them to the frames with the provided fasteners. Take care not to overdrive the fasteners.

Fasten the doorframe trim pieces to the frames to finish the door opening. If the fasteners are colored to match the trim, make sure you choose the correct ones.

(continued)

Building a Kit Shed (continued)

Insert the shed gable panels into the side frames and the door track and slide them together so the fastener holes are aligned. Attach the panels with the provided fasteners.

Fit the main roof beam into the clips or other fittings on the gable panels. Have a helper hold the free end of the beam. Position the beam and secure it to both gable ends before attaching it. Drive fasteners to affix the roof beam to the gable ends and install any supplementary support hardware for the beam, such as gussets or angle braces.

Begin installing the roof panels at one end, fastening them to the roof beam and to the top flanges of the side frames.

Apply weatherstripping tape to the top ends of the roof panels to seal the joints before you attach the overlapping roof panels. If your kit does not include weatherstripping tape, look for adhesive-backed foam tape in the weatherstripping products section of your local building center.

As the overlapping roof panels are installed and sealed, attach the roof cap sections at the roof ridge to cover the panel overlaps. Seal as directed. **Note:** Completing one section at a time allows you to access subsequent sections from below so that you don't risk damaging the roof.

Attach the peak caps to cover the openings at the ends of the roof cap and then install the roof trim pieces at the bottoms of the roof panels, tucking the flanges or tabs into the roof as directed. Install a plywood floor, according to manufacturer instructions.

Assemble the doors, paying close attention to right/left differences on double doors. Attach hinges for swinging doors and rollers for sliding doors.

Install door tracks and door roller hardware on the floor as directed and then install the doors according to the manufacturer's instructions. Test the action of the doors and make adjustments so the doors roll or swing smoothly and are aligned properly.

pole
barn

Pole structures have been a mainstay of American barn construction since the country's beginnings. Farmers originally built pole-framed buildings because of their economical use of lumber and their strength and durability. And, the materials were readily available: the poles used in early barns were usually tree trunks from elsewhere on the property that were felled to clear the land.

The main difference between pole barns and traditionally framed barns lies in the way the building's weight is distributed. In traditional frame construction, the building's weight rests on top of its foundation and a series of load-bearing walls or other structural elements. In pole construction, the building's weight hangs on the poles instead of resting on the foundation. This means that pole buildings can easily be quite large without a lot of foundation materials or interior supports. And, the builder can choose a variety of different flooring options, from bare earth to compacted gravel to poured concrete. Also, because the poles or posts are embedded deep in the ground (typically 4 feet or more), the ground can shift around them with seasonal freeze/thaw cycles without affecting the building at all. The builder also benefits from flexibility with regard to the interior of the barn—stalls for animals, walls, or storage space can easily be added, or the interior can be left completely open to make space to store large pieces of machinery or other large projects.

Today, pole structures are always built on poles or posts that have been chemically treated to protect the wood from rot and to ensure the long-term life of the building. Often, any framing elements that are attached near the ground—such as the skirt board in this project—are chemically treated as well. The barn featured here utilizes 29-gauge steel siding and roofing for a sturdy, industrial-quality building, but these materials can easily be substituted for cedar board-and-batten or tongue-and-groove siding and traditional roof sheathing and asphalt shingles to match other structures on your property.

Pole barns allow for great design flexibility, from the choice of finishing materials to the building's actual purpose and design. This version features a partial lean-to overhang on one side that could be used to temporarily shelter vehicles or to stack firewood and building materials.

309

TOOLS & MATERIALS

Excavation tools

Stakes

Mason's string

Hammer

Tape measure

Marking paint

Power skid steer with 22"-dia.
   auger dig-in attachment

Hand tamper

Maul

Shovel

Chainsaw

Reciprocating saw

Circular saw

4' level

Chalk line

Laser level on a tripod

Combination square

Framing square

Cordless drill

Tin snips or nibbler

Eye and ear protection

Work gloves

CUTTING LIST

DESCRIPTION	QTY/SIZE	MATERIAL
FOUNDATION		
Precast concrete footing pad	27@20"-dia × 6"-thick	
Eave post	16@16'	3-ply 2 × 6, treated on one end
Lean-to post	4@14'	3-ply 2 × 6, treated
Gable post 1	1@18'	3-ply 2 × 6, treated on one end
Gable post 2	3@20'	3-ply 2 × 6, treated on one end
Gable post 3	3@22'	3-ply 2 × 6, treated on one end
Uplift blocks	54@12"	treated 2 × 4
FLOOR		
Foundation base (optional)	31 cu. yds.	Compactable stone gravel
WALL FRAMING		
Skirt board	8@16'	Treated 2 × 6
Ribbon board	7@16'	2 × 4
Wall framing members	2@10', 6@12', 48@16', 12@18'	2 × 4
Truss spacer	14@8'	2 × 2
Lean-to eave nailers	24'	2 × 4
Corner bracing	16@16'	2 × 4
ROOF FRAMING		
Trusses	8 Preassembled, engineered for your building's design	
Purlins	84@18'	2 × 4
Lateral truss bracing	28@18'	2 × 4
Lean-to ledger	24'	2 × 10
Lean-to rafter	4@10'	2 × 10
Lean-to 90° return board	2@102"	2 × 6
Knee braces	2@12'	2 × 6
EXTERIOR FINISHES		
Tails	8@12'	2 × 4
Subfascia	13@18'	2 × 8
Eave molding trim	12@10'6"	29-gauge steel
Soffit starter	19@10'6"	1¾" 29-gauge steel
Base trim	18@10'6"	29-gauge steel
Eave wall siding	38@136"-long	29-gauge steel, 36"-wide panels
Gable wall siding	4@218¼", 4@206¼", 4@194¼", 4@182¼", 4@170¼", 4@158¼"	29-gauge steel, 36"-wide panels 36"-wide panels
Wall corner trim	4@14'6"	29-gauge steel
Outside roof corner trim	6@14'6"	29-guage steel
Vented soffit	140'	12 × 16" panel
Fascia	20@10'6"	5½" angle trim
Door jamb trim	2@10'6", 1@12'6"	7½" 29-gauge steel
J-channel trim	6@10'6"	29-gauge steel
ROOFING		
Roofing	40@239"-long	29-gauge steel, 36"-wide panels
Ridge cap	6@10'6"	20" metal ridge
WINDOWS & DOORS		
Windows	3	3' × 4'
Service doors	2	
Garage door		12' × 10'
HARDWARE		
4" steel pole barn nails	30 lbs	
6" steel pole barn nails	20 lbs	
16d common nails	20 lbs	
PVC Nails	2 lbs	
2" stainless steel screws	5 lbs	
Joist hangers	4	
Joist hanger nails		
1½" galvanized trim nails	5 lbs	
1½" gasket washer ell-cap screws	3000	

Floor Plan

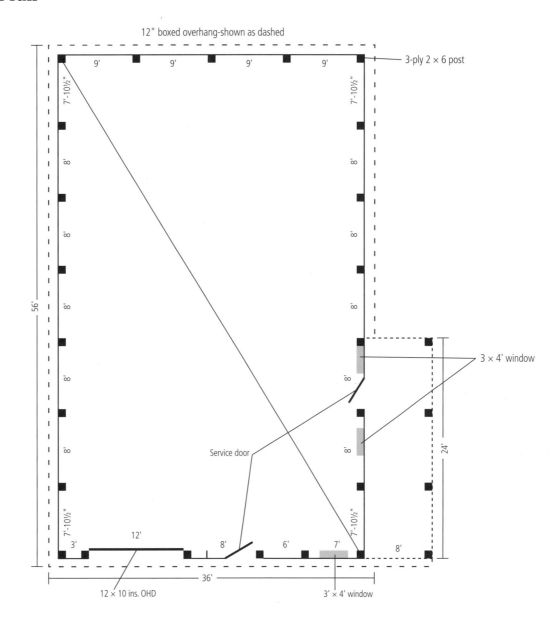

12" boxed overhang-shown as dashed

3-ply 2 × 6 post

9' 9' 9' 9'

7'-10½"

8'

8'

8'

8'

8'

7'-10½"

56'

7'-10½"

8'

8'

8'

8'

8'

7'-10½"

3 × 4' window

24'

Service door

3' 12' 8' 6' 7' 8'

36'

12 × 10 ins. OHD

3' × 4' window

Front Elevation

12
4

5½"-wide steel fascia

Boxed overhang

Steel corner trim

12' × 10' garage door

Service door

36"-wide steel siding panels

3-ply 2 × 6 treated post

3 × 4' window

36' wide 11'4" floor to ceiling

(continued)

311

Rear Elevation

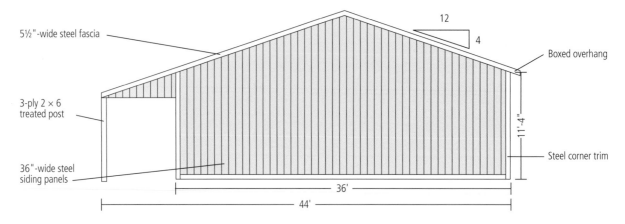

5½"-wide steel fascia

3-ply 2 × 6 treated post

36"-wide steel siding panels

12

4

Boxed overhang

Steel corner trim

11'-4"

36'

44'

Right Side Elevation

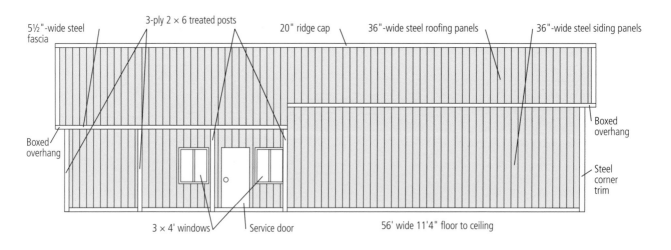

5½"-wide steel fascia

3-ply 2 × 6 treated posts

20" ridge cap

36"-wide steel roofing panels

36"-wide steel siding panels

Boxed overhang

Boxed overhang

Steel corner trim

3 × 4' windows

Service door

56' wide 11'4" floor to ceiling

Left Side Elevation

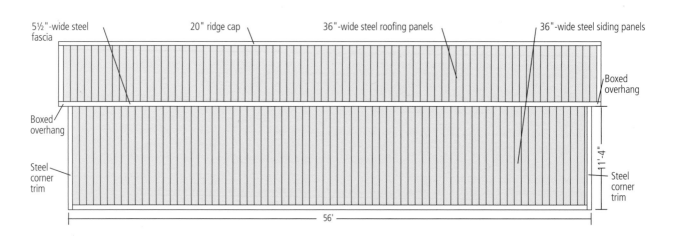

5½"-wide steel fascia

20" ridge cap

36"-wide steel roofing panels

36"-wide steel siding panels

Boxed overhang

Boxed overhang

Steel corner trim

11'-4"

Steel corner trim

56'

Left Side Framing Plan

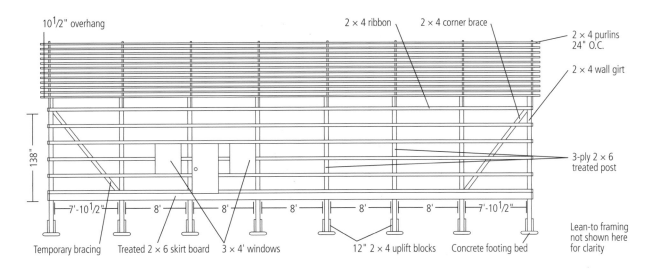

10 1/2" overhang · 2 × 4 ribbon · 2 × 4 corner brace · 2 × 4 purlins 24" O.C. · 2 × 4 wall girt · 3-ply 2 × 6 treated post · 138" · 7'-10 1/2" · 8' · 8' · 8' · 8' · 8' · 7'-10 1/2" · Temporary bracing · Treated 2 × 6 skirt board · 3 × 4' windows · 12" 2 × 4 uplift blocks · Concrete footing bed · Lean-to framing not shown here for clarity

Right Side Framing Plan

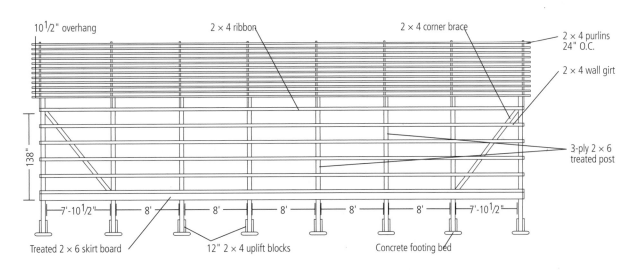

10 1/2" overhang · 2 × 4 ribbon · 2 × 4 corner brace · 2 × 4 purlins 24" O.C. · 2 × 4 wall girt · 3-ply 2 × 6 treated post · 138" · 7'-10 1/2" · 8' · 8' · 8' · 8' · 8' · 7'-10 1/2" · Treated 2 × 6 skirt board · 12" 2 × 4 uplift blocks · Concrete footing bed

Lean-to Framing Plan

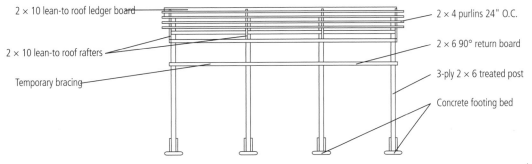

2 × 10 lean-to roof ledger board · 2 × 4 purlins 24" O.C. · 2 × 10 lean-to roof rafters · 2 × 6 90° return board · Temporary bracing · 3-ply 2 × 6 treated post · Concrete footing bed

(continued)

313

Front Framing Plan

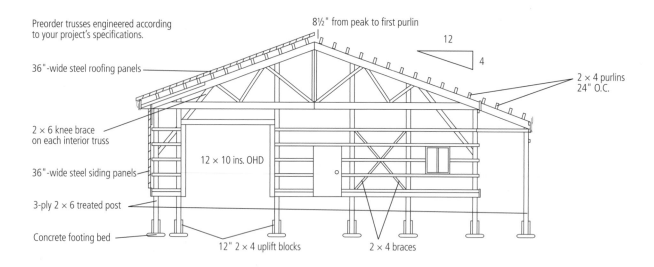

Preorder trusses engineered according to your project's specifications.

8½" from peak to first purlin

12
4

36"-wide steel roofing panels

2 × 4 purlins 24" O.C.

2 × 6 knee brace on each interior truss

36"-wide steel siding panels

12 × 10 ins. OHD

3-ply 2 × 6 treated post

Concrete footing bed

12" 2 × 4 uplift blocks

2 × 4 braces

Rear Framing Plan

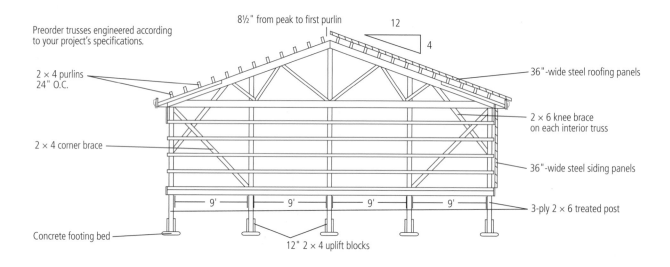

Preorder trusses engineered according to your project's specifications.

8½" from peak to first purlin

12
4

2 × 4 purlins 24" O.C.

36"-wide steel roofing panels

2 × 6 knee brace on each interior truss

2 × 4 corner brace

36"-wide steel siding panels

9'   9'   9'   9'

3-ply 2 × 6 treated post

Concrete footing bed

12" 2 × 4 uplift blocks

Lateral Truss Bracing Detail

The truss shown here is an example. Always order trusses engineered according to your project's specifications.

Your truss specifications sheet will indicate where lateral bracing should be placed, as in this example. Always follow the bracing recommendations for your truss design.

Lateral truss bracings

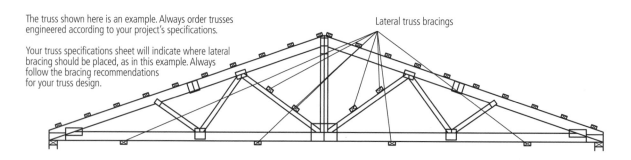

How to Build a Pole Barn with Partial Lean-to Overhang

Excavate the barn site approximately 3' longer and wider than the building's footprint (44 × 56' as shown in this project). Make sure the area is flat and all surrounding growth is removed. If you prefer to work on a gravel base, add a 4"-layer of compactable gravel now, or you can leave the excavated area bare. The foundation for this barn will be a perimeter of posts embedded in concrete beds within 4' footings.

Stake out the four corner post locations to form a rectangle. Posts should be positioned 1½" inside the building's actual footprint to leave room for the wall girts, which will be installed on the outside of the posts.

Warning

The following photos feature construction professionals who are very experienced working at dangerous heights. Do not attempt these techniques yourself. For your safety, always use ladders or scaffolding to access the upper work areas.

Set up the batter boards at least 12" outside of each corner. Pound a nail partially into the top of each batter board so that the strings intersect at the stake's location; tie a mason's string to one of these nails and run to the location of the next corner post. Wrap the string around this nail to form a line for the first wall. Then, wrap the string around the perpendicular batter board nail and run to the next corner.

Adjust batter boards, stakes, and string for square, as necessary. Finish by tying the end of the string to the second batter board outside the first corner post. Lay out post locations for lean-to corners according to the FLOOR PLAN.

(continued)

How to Build a Pole Barn with Partial Lean-to Overhang (continued)

Check the mason's string layout for level. Mount a laser level on a tripod in one location where you can easily shoot all four corner post locations. Hold a tape measure running up from the mason's strings along with the laser level receiver; find the laser level beam on the receiver and check that it shoots each corner at the same measurement on the tape measure. Adjust strings until the measurements in all four corners are equal.

Measure diagonally from corner to corner of the building outline to confirm square. After adjusting all measurements, confirm the measurements one last time by repeating steps 5 and 6. Setting up a properly leveled and square foundation is key to the success of the building. Unload the 2 × 6 skirt boards and place them around the perimeter of the building.

Mark wall post locations. Start by marking all corner posts on the ground with marking paint at the intersection of the mason's strings. Measure out with a tape measure from the first corner and mark post location. Continue marking post locations around the perimeter according to FLOOR PLAN. After post locations are marked, double-check corners for square and check that post locations on the eave sides are marked directly across from one another for proper alignment. Adjust and remark as needed.

22"-dia. auger

Dig post footings. In each corner, measure the distance from the mason's strings to the ground. Add the required post hole depth: here, 54" (48" for the post footing, plus 6" for the concrete footing bed) to the smallest of these measurements. Use a rented power skid steer with a 22"-dia. auger dig-in attachment to dig holes for the post footings to this depth.

Tamp post holes with a hand tamper or make your own tamper by attaching scrap lumber to the end of a long 2 × 4, as shown. Tamp hole bottoms, measure for depth, and check footing base for level multiple times. When the footing base is solid and level and the hole measures the correct depth, insert a precast, circular concrete footing bed into the bottom of each hole (inset).

Working Around Rocks

As you dig with an auger, you may encounter large stones, roots, compacted clay, or other hard ground material. When you confront this, it may be necessary to use a maul or jack hammer to break up the material and then shovel out the loosened soil and/or rocks.

Uplift blocks

Center 12"-long 2 × 4 uplift blocks on two opposing sides of the treated ends of the posts, flush with the bottom ends. Attach with two to three 16d galvanized common nails. After backfilling footings, these blocks will help to anchor the post in the ground and protect it from shifting due to weather or other environmental changes. Cut the posts to the correct length.

Drop all posts into holes. Drop the chemically treated end of each post into the footing. Rest the post in the footing so the end leans in toward the building's interior. Double-check post location measurements and make sure mason's strings are level before proceeding to step 12. This is also a good time to unload purlins, wall girts, and temporary bracing lumber around your work site.

(continued)

How to Build a Pole Barn with Partial Lean-to Overhang (continued)

Raise the first corner post. Set the wall posts so the 4½"-side faces the exterior of the building (the corner posts should be positioned so the 4½"-side faces the eave wall). Use a 4' level to check for plumb.

Backfill around the plumb post with the soil you removed to dig the footings. As you hold the post, maintaining plumb, have one or two helpers backfill and tamp soil firmly. When set, measure to the next post to ensure accurate placement. Continue around the structure and lean-to until all posts have been plumbed and backfilled.

Mark level reference lines. Set a laser level on a tripod inside the building footprint. Set the level's height and shoot one post at a time; use a combination square to make a mark on each post. Be careful not to jostle the level or change its height or placement at all during this process. Then, measure down on each post from this mark to the ground and record the measurement. The smallest measurement will determine the placement of the skirt board around the structure.

Attach the skirt board. Mark the measurement from step 14 onto the exterior face of the two adjacent corner posts and pound a nail about halfway in at the mark. Connect the nails on the outside of the posts with mason's strings (as shown). Pound a nail halfway into the exterior face of each wall post between the corners at the string's location. Align the first 2 × 6 skirt board with its bottom edge flush to the placement nails and attach to posts with two 4" steel pole barn nails. Continue installing skirt boards to finish the wall. **NOTE:** On eave walls, position the board so its end extends beyond each corner post 1½". On gable walls, cut the skirt board to fit flush to the edge of the post, creating a 1½" corner.

Attach the bottom 16' 2 × 4 wall girt nailer directly above the skirt board. Attach to each post with two 4" steel pole barn nails.

Attach the four remaining wall girt nailers every 24" on-center up the posts. Each wall girt will create a 1½" corner, like the skirt board. Repeat steps 15 to 17 to install skirt boards and wall girts on the opposite eave wall. Attach a temporary wall girt to the outside of the lean-to posts to keep them square during construction.

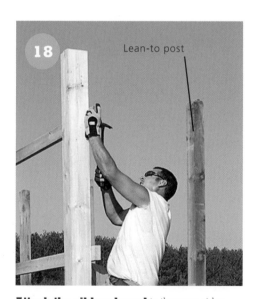

Attach the ribbon board to the eave sides. Hook your measuring tape to the bottom of the skirt board and measure up to 138". Mark this location on all eave-side posts. Align a 2 × 4 ribbon board with its bottom edge flush to this mark and attach to each post with two 4" steel pole barn nails. Note: Always use ladders and safety equipment. The worker pictured here is a professional.

Install 2 × 4 corner bracing between the corner post and the adjacent wall post, from the bottom of one post to the ribbon of the next. Nail bracing to wall girts.

(continued)

How to Build a Pole Barn with Partial Lean-to Overhang (continued)

Notch the top of each eave wall post to support the trusses. Cut notches to the top of the ribbon. Then, measure up from the ribbon board to the height of the truss heel and make a mark. Starting at this mark, cut the top of the post to match the truss's ⁴⁄₁₂ pitch.

Prepare trusses for installation. Unload preassembled trusses designed to meet your building's specifications and approved by an engineer for use with your building. Before installation, mark the location of the first purlins on each side of the peak at 8½".

Raise the first truss. Place the first truss on the outside of the gable-wall posts, with the bottom edge of the bottom chord flush with the top edge of the ribbon board on the eave sides. Check for level and nail the truss to all posts with four or five 4" steel pole barn nails at each post location. **OPTION:** Equip a skid steer with a lift to help raise each truss to the top of the posts (below), where one person on each side of the building should be waiting to guide it into place.

Raise the second truss to rest in the notches of the second set of posts. Check for level and nail to the remaining plies of the posts with four or five 4" steel pole barn nails. The notched end of the posts should extend up on one side of the truss.

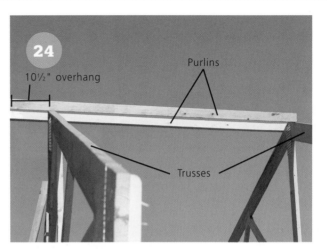

Tie the trusses together with purlins. Before you disconnect the second truss from the lift, position the first 2 × 4 purlin on-edge with its bottom edge flush to the mark made in step 21 and its end extended 10½" past the gable side of the truss. Nail the purlin to the truss with one 6" steel pole barn nail; repeat with a second purlin on the opposite side of the peak. Attach the first two purlins to the second truss in the same manner and disconnect the second truss from the lift.

Install the remaining trusses, repeating steps 23 and 24 for each truss. A section of three trusses makes up a bay. After each bay is completed, slide a new purlin down the length of the previous one for the next bay. Although they do not need to be fastened together, overlap the purlins approximately 12" between bays. When completed, the purlins will appear to be staggered slightly because of this overlap.

Trim posts on the gable end to follow the ⁴/₁₂ pitch of the trusses using a chainsaw. Install skirt boards, wall girts, and 2 × 4 corner bracing on the gable walls at this time (see steps 15 to 17 and 19). If a door or window opening impedes on the corner space for the bracing, attach a brace in the next available opening between posts.

(continued)

How to Build a Pole Barn with Partial Lean-to Overhang (continued)

Attach the remaining purlins on both sides of the peak, spaced every 24" on-center down the length of the roof. All purlins should extend 10½" past both gable ends and should be installed to cover a bay. This building has one extra truss-to-truss space not included in a bay; select a space after a full bay and use 9' purlins for this space, overlapping the purlins on either side by 6".

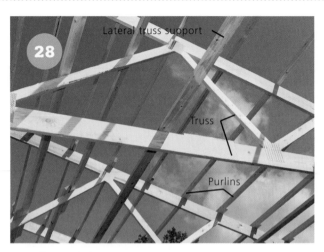

Install lateral truss support according to your truss design's specification sheet. The trusses used in this project required two 2 × 4 braces on the webbing and four 2 × 4 braces running on top of the bottom chord (pictured here, see LATERAL TRUSS DETAIL). Cut 2 × 4 bracing boards to 18-ft. and overlap them about 12" after each bay. Attach to trusses with 4" steel pole barn nails.

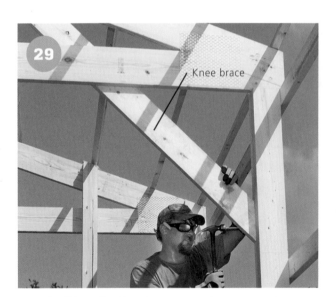

Install 12'-long 2 × 6 knee bracing to each interior truss/post to keep the walls straight during roofing material installation. Cut a 45° angle into each end of the brace and install so its length is split equally between the truss and the inside of the building; attach the top 6' to the truss, and attach the bottom of the brace to the corresponding interior post with 4" steel pole barn nails. Repeat on the other side.

Install the ledger board for the lean-to rafters. Cut a 2 × 10 and attach it flush with the top chords of the trusses. Nail to the ends of the trusses with four 4" steel pole barn nails at each location. Notch 1½" from the top of the exterior lean-to posts as in step 20 to a depth of 8".

Fasten the lean-to rafters to the ledger board using joist hangers. Use a framing square to mark one end of each of the four 2 × 10 rafters to match the roof's ½₂ pitch. Cut this angle. Attach the angled end to the ledger using joist hangers and joist hanger nails, and attach the outer end to the notched portion of the lean-to post with 4" steel pole barn nails. Cut the end of the rafter at the same ½₂ pitch 1½" past the outer edge of the post.

Attach a 90° return board to the outside of the posts at both gable ends of the lean-to roof, even with the top wall girt on the building. Nail the return boards to the posts with 4" steel pole barn nails. Cut the return boards to be flush with the outside of the posts. Install a 2 × 4 ribbon board on the eave side of the lean-to flush with the 90° return boards with two 4" steel pole barn nails.

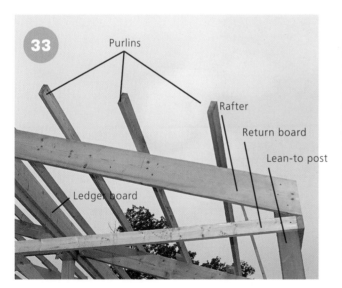

Install purlins on the lean-to rafters, staying consistent with the purlin spacing on the roof. Position the first lean-to roof purlin 24" on-center from the last purlin on the building roof, and attach it with one 6" steel pole barn nail at each rafter. Continue until all purlins on the lean-to roof are installed.

Cut openings for service doors, windows, and a garage door according to these components' specifications. Install service doors and windows as directed. The garage door will be installed later. Frame the openings.

(continued)

How to Build a Pole Barn with Partial Lean-to Overhang (continued)

Cut and attach the first 2 × 4 tail to a gable wall truss to create the boxed overhang. Use a framing square to measure and cut one end of a 2 × 4 to match the roof's ¹⁄₁₂ pitch. Set the tail so the angled end is 10½" from the wall; a few feet of 2 × 4 will extend along the truss chords. Attach the tail to the truss with five 16d common nails, spaced about 10" apart. Next, install the tail on the opposite gable wall truss on the other end of the wall in the same manner.

To install the tails on the rest of the wall, pound a 16d common nail halfway into the end of both the first and last tails and connect the nails with a mason's string spanning from one end of the building to the other. Install the remaining tails flush to this string. Install tails on the opposite side of the building in this same manner.

Install 2 × 4 spacers between the rafters on the eave edge of the lean-to roof (these are nailers for subfascia). Mark a consistent spacer location on each rafter end and then measure and cut 2 × 4s to fit between the rafters. Attach the spacers to the rafters and posts with 4" pole barn nails. Before you install the subfascia, install temporary 2 × 4 bracing for each truss; wedge one end where the truss meets the post and rest the other on the ground to create a firm brace.

Attach 2 × 8 subfascia to the outside of the truss tails with 4" steel pole barn nails. Attach the first piece on the eave side with its bottom edge flush to the bottom edge of the tail, flush with the gabled end. Measure and cut the subfascia as you go. Cut and position the subfascia so that each board covers only half the tail face and the next piece can be easily attached.

Install 2 × 2 spacers on top of the ribbon between each truss. Measure each opening and cut the 2 × 2 to fit, then fasten it to the top of the ribbon with pole barn nails.

Install the soffit starter. This two-in-one trim has a J-channel side to cover the top of the wall and a C-channel side to receive the soffit. Mark the location of the soffit starter by marking a level line on the framing members of the building, leveling over from the bottom of the subfascia. Attach to the ribbons and 2 × 2 spacers with stainless-steel screws on all four walls and on the front side of the lean-to (the back gable and eave side of the lean-to do not have soffits).

Install base trim (left photo) to the 2 × 6 skirt board, with its bottom edge flush with the joint between the skirt board and the first 2 × 4 wall girt. Install door jamb trim (right photo) and J-channel around the garage door opening and install J-channel around the service door. Install all doors and windows.

(continued)

How to Build a Pole Barn with Partial Lean-to Overhang (continued)

Install the steel siding. Steel wall siding is made up of precut 36"-wide panels with 9"-on-center ribs. Attach the siding flush to the first corner, and hold it to the wall girts temporarily with two or three 1½" gasket washer bell cap screws per panel. Overlap the first rib of the next sheet with the last rib of the previous sheet and tack it in place. Cut steel panels to fit around door and window openings with a tin snips or a nibbler (right photo). Remember to side the gable ends of the lean-to roof as well. Finish installing siding around the building.

Install soffit panels and eave molding trim. Fit soffit panels into the C-channel and attach the other end to the bottom of the subfascia (left photo). Install steel eave molding trim to cover the top of the subfascia and slip it under the roof seal on the eave sides of the roof using galvanized trim nails (right photo).

Install the steel roof. Order steel to the correct dimensions for your project; for this building, steel panels are 36" wide. Begin on one gable side, and attach steel roofing to each purlin in the flats with 1½" gasket washer bell cap screws. Position the first piece so you have an overhang of between 2 to 3". Keep this overhang consistent for the remainder of the roofing installation. Overlap the ridge of each panel over the previous panel as you move down the roof. On the side of the roof with the lean-to, start installing the roofing on the lean-to roof, and then move up to the building roof, overlapping panels as you go. When you get to the end, cut the last piece of roofing to fit.

Install the ridge cap. First, install the closure between the roof steel and the ridge cap, composite material that allows for excellent airflow, but prevents dust and debris from entering the building. One side of this porous, thick material will be pre-glued; its profile will match the steel roofing material. Set the ridge cap on top of the composite material and attach with 2" stainless-steel screws at each purlin.

Complete the siding installation by adding the remaining screws. Use a chalk line to mark a long line onto the siding at each wall girt location from one end of the building to the other. Use this chalk line to drive screws into the 9"-on-center flats in the siding, one at each wall girt location.

Attach finishing trim. Install corner trim on all corners with 2" stainless-steel screws. Attach fascia to cover the edge of the soffit and the remaining exposed subfascia with 2" stainless-steel screws. Remove temporary bracing. Fill the gap beneath the skirt board around the building with excess soil, if desired. **OPTION:** Finish this building with a concrete floor. Unless you are experienced working with concrete, call in the professionals to finish a floor of this size.

post-and-board fence

45

Post-and-board fences include an endless variety of simple designs in which widely spaced square or round posts support several horizontal boards. This type of fence has been around since the early 1700s, remaining popular for its efficient use of lumber and land and its refined appearance.

The post-and-board is still a great design today. Even in a contemporary suburban setting, a classic, white three- or four-board fence evokes the stately elegance of a horse farm or the welcoming, down-home feel of a farmhouse fence bordering a country lane.

Another desirable quality of post-and-board fencing is its ease in conforming to slopes and rolling ground. In fact, it often looks best when the fence rises and dips with ground contours. Of course, you can also build the fence so it's level across the top by trimming the posts along a level line. Traditional agricultural versions of post-and-board fences typically include three to five boards spaced evenly apart or as needed to contain livestock. If you like the look of widely spaced boards but need a more complete barrier for pets, cover the back side of the fence with galvanized wire fencing, which is relatively unnoticeable behind the bold lines of the fence boards. You can also use the basic post-and-board structure to create any number of custom designs. The fence styles shown in the following pages are just a sampling of what you can build using the basic construction technique for post-and-board fences.

TOOLS & MATERIALS

Mason's string
Line level
Circular saw
Speed square
Clamps
4 × 4 posts
Finishing materials
Drill
Chisel
Concrete

Primer paint or stain
3" stainless-steel screws
Post levels
Combination square
Lumber (1 × 6, 1 × 4, 2 × 6, 1 × 3)
Deck screws 2", 2½", and 3½"
8d galvanized nails
Work gloves
Pencil
Eye and ear protection

A low post-and-board fence, like traditional picket fencing, is both decorative and functional, creating a modest enclosure without blocking views. The same basic fence made taller and with tighter board spacing becomes an attractive privacy screen or security fence.

Building a Classic Post-and-Board Fence

Set the posts in concrete, following the desired spacing. Laying out the posts at 96" on center allows for efficient use of lumber. For smaller boards, such as 1 × 4s and smaller, set posts closer together for better rigidity.

Trim and shape the posts with a circular saw. For a contoured fence, measure up from the ground and mark the post height according to your plan (post height shown here is 36"). For a level fence, mark the post heights with a level string. If desired, cut a 45° chamfer on the post tops using a speed square to ensure straight cuts. Prime and paint (or stain and seal) the posts.

Mark the board locations by measuring down from the top of each post and making a mark representing the top edge of each board. The traditional 3-board design employs even spacing between boards. Use a speed square to draw a line across the front faces of the posts at each height mark. Mark the post centers on alternate posts using a combination square or speed square and pencil. For strength, it's best to stagger the boards so that butted end joints occur at every other post (this requires 16-ft. boards for posts set 8-ft. apart). The centerlines represent the location of each butted joint.

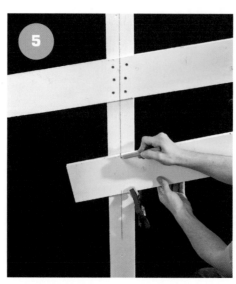

Install 1 × 6 boards. Measure and mark each board for length, and then cut it to size. Clamp the board to the posts, following the height and center marks. Drill pilot holes and fasten each board end with three 2½" deck screws or 8d galvanized box nails. Use three fasteners where long boards pass over posts as well.

Mark for mitered butt joints at changes in elevation. To mark the miters on contoured fences, draw long centerlines onto the posts. Position an uncut board over the posts at the proper height, and then mark where the top and bottom edges meet the centerline. Connect the marks to create the cutting line, and make the cut. **Note:** The mating board must have the same angle for a symmetrical joint.

Building a Notched-Post Fence

1 × 4 board

6"

9½"

15½"

19"

25"

28½"

7½" gap

Side View–Post

Post

¾"
deep
notches

The notched-post fence presents a slight variation on the standard face-mounted fence design. Here, each run of boards is let into a notch in the posts so the boards install flush with the post faces. This design offers a cleaner look and adds strength overall to the fence. In this example, the boards are 1 × 4s so the posts are set 6' on center; 1 × 6 or 2 × 6 boards would allow for wider spacing (8'). **Note:** Because the notches must be precisely aligned between posts, the posts are set and braced before the concrete is added. Alternatively, you can complete the post installation and then mark the notches with a string and cut each one with the posts in place.

Cut and mark the posts. Cut the 4 × 4 posts to length at 66". Clamp the posts together with their ends aligned, and mark the notches at 6, 9½, 15½, 19, 25, and 28½" down from the top ends.

Create the notches. Make a series of parallel cuts between the notch marks using a circular saw with the blade depth set at ¾". Clean out the waste and smooth the bases of the notches with a chisel.

Install the posts and boards. Set the posts in their holes and brace them in place using a level string to align the notches. Secure the posts with concrete. Prefinish all fence parts. Install the 1 × 4 boards with 2" deck screws (driven through pilot holes) so their ends meet at the middle of each post.

Building a Capped Post-and-Board Fence

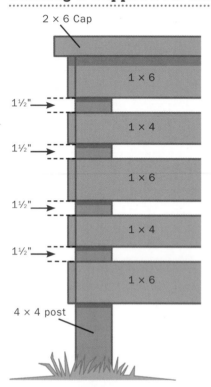

2 × 6 Cap

1 × 6

1½"

1 × 4

1½"

1 × 6

1½"

1 × 4

1½"

1 × 6

4 × 4 post

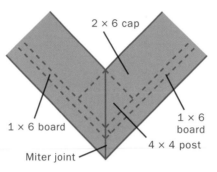

2 × 6 cap

1 × 6 board

Miter joint

1 × 6 board

4 × 4 post

Top View–Detail

A cap rail adds a finished look to a low post-and-board fence. This fence design includes a 2 × 6 cap rail and an infill made of alternating 1 × 4 and 1 × 6 boards for a decorative pattern and a somewhat more enclosed feel than you get with a basic 3-board fence. The cap pieces are mitered over the corner posts. Where cap boards are joined together over long runs of fence, they should meet at a scarf joint—made with opposing 30 or 45° bevels cut into the end of each board. All scarf and miter joints should occur over the center of a post.

Install and mark the posts. Set the 4 × 4 posts in concrete with 72" on-center spacing. Trim the post tops so they are level with one another and approximately 36" above grade. Prefinish all fence parts. Use a square and pencil to mark a vertical centerline on each post where the board ends will butt together.

Install the boards. For each infill bay, cut two 1 × 4s and three 1 × 6s to length. Working from the top of the posts down, fasten the boards with 2½" deck screws driven through pilot holes. Use a 1½"-thick spacer (such as a 2 × 4 laid flat) to ensure even spacing between boards.

Add the cap rail. Cut the cap boards so they will install flush with the inside faces and corners of the posts; this creates a 1¼" overhang beyond the boards on the front side of the fence. Fasten the cap pieces to the posts with 3½" deck screws driven through pilot holes.

Building a Modern Post-and-Board Privacy Fence

This beautiful, modern-style post-and-board fence is made with pressure-treated 4 × 4 posts and clear cedar 1 × 3, 1 × 4, and 1 × 6 boards. To ensure quality and color consistency, it's a good idea to hand-pick the lumber, and choose S4S (surfaced on four sides) for a smooth, sleek look. Alternative materials include clear redwood, ipé, and other rot-resistant species. A high-quality, UV-resistant finish is critical to preserve the wood's natural coloring for as long as possible.

Install the posts, spacing them 60" on-center or as desired. Mark the tops of the posts with a level line, and trim them at 72" above grade. **Note:** This fence design is best suited to level ground. Cut the fence boards to length. If desired, you can rip down wider stock for custom board widths (but you'll have to sand off any saw marks for a finished look).

Fasten the boards to the post faces using 2½" deck screws or 8d galvanized box nails driven through pilot holes. Work from the top down, and use ⅞"-thick wood spacers to ensure accurate spacing.

Add the battens to cover the board ends and hide the posts. Use 1 × 4 boards for the infill posts and 1 × 6s for the corner posts. Rip ¾" from the edge of one corner batten so the assembly is the same width on both sides. Fasten the battens to the posts with 3" stainless-steel screws (other screw materials can discolor the wood).

split-rail fence

46

The split-rail, or post-and-rail, fence is essentially a rustic version of the post-and-board fence style (pages 328 to 333) and is similarly a good choice for a decorative accent, for delineating areas, or for marking boundaries without creating a solid visual barrier. Typically made from split cedar logs, the fence materials have naturally random shaping and dimensions, with imperfect details and character marks that give the wood an appealing hand-hewn look. Natural weathering of the untreated wood only enhances the fence's rustic beauty.

The construction of a split-rail fence couldn't be simpler. The posts have holes or notches (called mortises) cut into one or two facets. The fence rails have trimmed ends (called tenons) that fit into the mortises. No fasteners are needed. Posts come in three types to accommodate any basic configuration: common posts have through mortises, end posts have half-depth mortises on one facet, and corner posts have half-depth mortises on two adjacent facets. The two standard fence styles are two-rail, which stand about 3 feet tall, and three-rail, which stand about 4 feet tall. Rails are commonly available in 8- and 10-feet lengths.

In keeping with the rustic simplicity of the fence design, split-rail fences are typically installed by setting the posts with tamped soil and gravel instead of concrete footings (frost heave is generally not a concern with this fence, since the joints allow for plenty of movement). This comes with a few advantages: the postholes are relatively small, you save the expense of concrete, and it's much easier to replace a post if necessary. Plan to bury about a third of the total post length (or 24 inches minimum). This means a 3-foot-tall fence should have 60-inch-long posts. If you can't find long posts at your local home center, try a lumberyard or fencing supplier.

A split-rail fence looks great as a garden backdrop or a friendly boundary line. The rough-hewn texture and traditional wood joints are reminiscent of homesteaders' fences built from lumber cut and dressed right on the property.

How to Build a Split-Rail Fence

TOOLS & MATERIALS

Mason's string
Shovel
Clamshell digger or power auger
Digging bar (with tamping head) or 2 × 4
Level
Reciprocating saw or handsaw
Tape measure
Stakes
Soil
Nails
Precut split-rail fence posts and rails
Compactable gravel (bank gravel or pea gravel)
Plastic tags
Lumber and screws for cross bracing
Wheelbarrow
Eye and ear protection
Work gloves

Determine the post spacing by dry-assembling a fence section and measuring the distance between the post centers. Be sure the posts are square to the rails before measuring.

Set up a string line using mason's string and stakes to establish the fence's path, including any corners and return sections. Mark each post location along the path using a nail and plastic tag.

Dig the postholes so they are twice as wide as the posts and at a depth equal to ⅓ the total post length plus 6". Because split posts vary in size, you might want to lay out the posts beforehand and dig each hole according to the post size.

Add 6" of drainage gravel to each posthole. Tamp the gravel thoroughly with a digging bar or a 2 × 4 so the layer is flat and level.

Set and measure the first post. Drop the post in its hole, and then hold it plumb while you measure from the ground to the desired height. If necessary, add or remove gravel and re-tamp to adjust the post height.

Brace the post with cross bracing so it is plumb. Add 2" of gravel around the bottom of the post. Tamp the gravel with a digging bar or 2 × 4, being careful not to disturb the post.

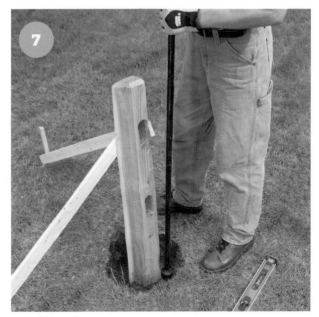

Fill and tamp around the post, one layer at a time. Alternate between 4" of soil and 2" of gravel (inset), tamping each layer all the way around the post before adding the next layer. Check the post for plumb as you work. Overfill the top of the hole with soil and tamp it into a hard mound to help shed water.

(continued)

How to Build a Split-Rail Fence (continued)

Assemble the first section of fence by setting the next post in its hole and checking its height. Fit the rails into the post mortises, and then brace the second post in place. **Note:** Set all the posts at the same height above grade for a contoured fence. For a level fence, follow a level mason's string from the top of the first post.

Secure the second post by filling and tamping with alternate layers of gravel and soil, as with the first post. Repeat steps 5 through 9 to complete the fence. **TIP:** Set up a mason's string to help keep the posts in a straight line as you set them.

Custom Details

Custom-cut your rails to build shorter fence sections. Cut the rails to length using a reciprocating saw and long wood blade or a handsaw (be sure to factor in the tenon when determining the overall length). To cut the tenon, make a cardboard template that matches the post mortises. Use the template to mark the tenon shape onto the rail end, and then cut the tenon to fit.

Gates for split-rail fences are available from fencing suppliers in standard and custom-order sizes. Standard sizes include 4' for a walk-through entrance gate and 8 or 10' for a drive-through gate. For large gates, set the side posts in concrete footings extending below the frost line.

Virginia Rail Fence

The Virginia rail fence—also called a worm, snake, or zigzag fence—was actually considered the national fence by the US Department of Agriculture prior to the advent of wire fences in the late 1800s. All states with farmland cleared from forests had them in abundance.

The simplest fences were built with an extreme zigzag and didn't require posts. To save on lumber and land, farmers began straightening the fences and burying pairs of posts at the rail junctures.

A variation in design that emerged with entirely straight lines is called a Kentucky rail fence.

Increase the zigzag to climb rolling ground or decrease it to stretch the fence out. For longevity, raise the bottom rail off the ground with stones or blocks. Posts will eventually rot belowground, but the inherently stable zigzag form will keep the fence standing until you can replace them.

Create layout lines equal to the amount of switchback on each section. Set the posts in the holes with gravel and bind the post pairs with rope. Install the split rails in alternating courses at each post pair, keeping the overhangs even.

Bind the tops of the posts together permanently with 9-gauge galvanized wire to hold the rails in position. Tighten the wire by twisting with a screwdriver blade as if you were tightening a tourniquet.

chain-link fence

If you're looking for long-lasting, economical fencing, a chain-link fence may be the perfect solution. Chain-link fences require minimal maintenance and provide excellent security. Erecting a chain-link fence is relatively easy, especially on level property. Leave contoured fence lines to the pros. For a chain-link fence with real architectural beauty, consider a California-style chain-link with wood posts and rails (see pages 346 to 347).

A 48-inch-tall chain-link fence—the most common choice for residential use—is what we've demonstrated here. The posts, fittings, and chain-link mesh, which are made from galvanized metal, can be purchased at home centers and fencing retailers. The end, corner, and gate posts, called terminal posts, bear the stress of the entire fence line. They're larger in diameter than line posts and require larger concrete footings. A footing three times the post diameter is sufficient for terminal posts. A properly installed stringer takes considerable stress off the end posts by holding the post tops apart.

When the framework is in place, the mesh must be tightened against it. This is done a section at a time with a winch tool called a come-along. As you tighten the come-along, the tension is distributed evenly across the entire length of the mesh, stretching it taut against the framework. One note of caution: it's surprisingly easy to topple the posts if you over-tighten the come-along. To avoid this problem, tighten just until the links of the mesh are difficult to squeeze together by hand.

Instructions for installing a chain-link gate are given on page 345. If you're building a new fence, it's a good idea to test-fit the gate to make sure the gate posts are set properly before you complete the fence assembly.

Chain-link fencing is a strong, durable, and inexpensive way to keep your crops safe from deer and racoons and keep pets and chickens and other farm animals confined.

TOOLS & MATERIALS

Supplies for
  setting posts
Mason's string
Ratchet wrench
Pliers
Hacksaw
  or pipe cutter
Chain-link fence
  materials
  and hardware
Duct tape
Tie wire
Circular saw,
  reciprocating saw, or
  handsaw
Drill
Come-along
  with spread bar
  and wire grip
Hog ring pliers
4 × 4 posts

2 × 4 lumber
3" deck screws or 16d
  galvanized common
  nails
Post finials or caps
Tension wire
Large galvanized
  fence staples
Hog rings
Lumber for
  cross bracing
Level
Permanent marker
Speed square
Clamps
Eye and ear protection
Hammer
Tape measure
Pencil
Work gloves

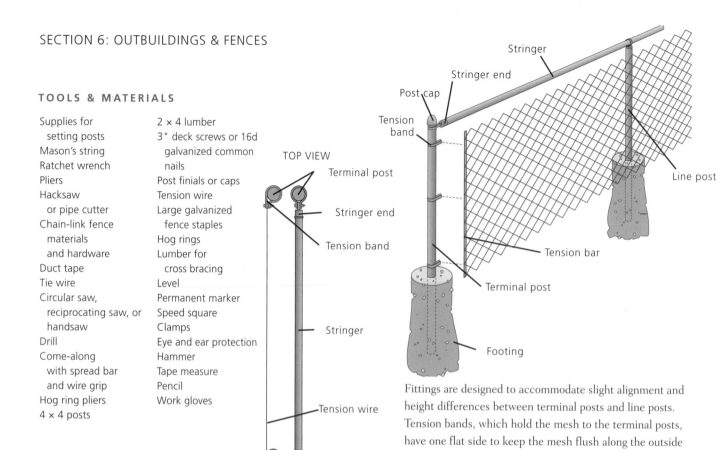

Fittings are designed to accommodate slight alignment and height differences between terminal posts and line posts. Tension bands, which hold the mesh to the terminal posts, have one flat side to keep the mesh flush along the outside of the fence line. The stringer ends hold the top stringer in place and keep it aligned. Loop caps on the line posts position the top stringer to brace the mesh.

How to Build a Chain-link Fence

Install the posts. Lay out the fenceline, spacing the posts at 96" on-center. Dig holes for terminal posts 8" in diameter with flared bottoms; dig holes for line posts at 6". Make all postholes 30" deep or below the frost line, whichever is deeper. Set the terminal posts in concrete so they extend 50" above grade. Run a mason's string between terminal posts at 46" above grade. Set the line posts in concrete so their tops are even with the string. If desired, stop the concrete 3" below ground level and backfill with soil and grass to conceal the concrete. **TIP:** When plumbing and bracing posts, use duct tape to secure cross bracing to the posts.

Position the tension bands and stringer ends on the gate and end terminal posts, using a ratchet wrench to tighten the bands with the included bolt and nut. Each post gets three tension bands: 8" from the top, 24" from the top, and 8" above the ground (plus a fourth band at the bottom of the post if you will use a tension wire). Make sure the flat side of each band faces the outside of the fence and points into the fence bay. Also add a stringer end to each post, 3" down from the top.

Add bands and ends to the corner posts. Each corner post gets six tension bands, two at each location: 8" and 24" from the top and 8" from the bottom (plus two more at the bottom for a tension wire, if applicable). Also install two stringer ends, 3" from the top of the post. Orient the angled side up on the lower stringer end and the angled side down on the upper stringer end.

Top each terminal post with a post cap and each line post with a loop cap. Make sure the loop cap openings are perpendicular to the fenceline, with the offset side facing the outside of the fenceline.

Begin installing the stringer, starting at a terminal post. Feed the non-tapered end of a stringer section through the loop cap on the nearest line post, then into the stringer end on the terminal post. Make sure it's snug in the stringer end cup. Continue feeding stringer sections through loop caps, and join stringer sections together by fitting the non-tapered ends over the tapered ends. If necessary, use a sleeve to join two non-tapered ends.

Measure and cut the last stringer section to fit to complete the stringer installation. Measure from where the taper begins on the preceding section to the end of the stringer end cup. Cut the stringer to length with a hacksaw or pipe cutter. Install the stringer.

(continued)

How to Build a Chain-link Fence (continued)

Secure the chain-link mesh to a terminal post, using a tension bar threaded through the end row of the mesh. Anchor the bar to the tension bands so the mesh extends about 1" above the stringer. The nuts on the tension bands should face inside the fence. If applicable, install a tension wire as directed by the manufacturer. Unroll the mesh to the next terminal post, pulling it taut as you go.

Stretch the mesh toward the terminal post using the come-along. Thread a spread bar through the mesh about 48" from the end, and attach the come-along between the bar and terminal post. Pull the mesh until it's difficult to squeeze the links together by hand. Insert a tension bar through the mesh and secure the bar to the tension bands. Remove excess mesh by unwinding a strand. Tie the mesh to the stringer and line posts every 12" using tie wire. See page 345 to install a gate.

Weaving Chain-link Mesh Together

If a section of chain-link mesh comes up short between the terminal posts, you can add another piece by weaving two sections together.

With the first section laid out along the fenceline, estimate how much more mesh is needed to reach the next terminal post. Overestimate 6" or so, so you don't come up short again.

Detach the amount of mesh needed from the new roll by bending back the knuckle ends of one zigzag strand in the mesh. Make sure the knuckles of the same strand are undone at the top and bottom of the fence. Spin the strand counter-clockwise to wind it out of the links, separating the mesh into two.

Place this new section of chain-link at the short end of the mesh so the zigzag patterns of the links line up with one another.

Weave the new section of chain-link into the other section by reversing the unwinding process. Hook the end of the strand into the first link of the first section. Spin the strand clockwise until it winds into the first link of the second section, and so on. When the strand has connected the two sections, bend both ends back into a knuckle. Attach the chain-link mesh to the fence framework.

How to Build a Chain-link Gate

Set fence posts in concrete spaced far enough apart to allow for the width of the gate plus required clearance for the latch. Position the female hinges on the gate frame, as far apart as possible. Secure with nuts and bolts (orient nuts toward the inside of the fence).

Set the gate on the ground in the gate opening, next to the gatepost. Mark the positions of the female hinges onto the gate post. Remove the gate and measure up 2" from each hinge mark on the gatepost. Make new reference marks for the male hinges.

Secure the bottom male hinge to the gatepost with nuts and bolts. Slide the gate onto the bottom hinge. Then, lock the gate in with the downward-facing top hinge.

Test the swing of the gate and adjust the hinge locations and orientations, if needed, until the gate operates smoothly and the opposite side of the gate frame is parallel to the other fence post. Tighten the hinge nuts securely.

Attach the gate latch to the free side of the gate frame, near the top of the frame. Test to make sure the latch and gate function correctly. If you need to relocate a post because the opening is too large or too small, choose the latch post, not the gate post.

345

How to Build a California-style Chain-link Fence

Install the posts. Set the 4 × 4 fence posts in concrete, spacing them at 6 to 8' on center. The posts should stand at least 4" taller than the finished height of the chain-link mesh.

Trim the posts so they are 4" higher than the installed height of the chain-link mesh. Mark the post height on all four sides of each post, and make the cuts with a circular saw, reciprocating saw, or handsaw.

Add 2 × 4 top stringers between each pair of posts. Mark reference lines 4" down from the tops of the posts. Cut each stringer to fit snugly between the posts. Fasten the stringers with their top faces on the lines using 3" deck screws or 16d galvanized common nails driven through angled pilot holes.

Wrap tension wire around a terminal post, about 1" above the ground. Staple the wire with a galvanized fence staple, and then double back the tail of the wire and staple it to the post.

Staple the tension wire to the line posts after gently tightening the wire (using a come-along with a wire grip) and securing the loose end of the wire to the opposing terminal post. **OPTION:** You can install 2 × 4 bottom stringers in place of a tension wire.

Add finials or decorative caps to the post tops for a finished look and to help protect the end grain of the wood.

Secure the fence mesh to the first terminal post using a tension bar threaded through the end row of the mesh. Fasten the bar to the posts with a fence staple every 8". Make sure the bar is plumb and the top of the mesh overlaps the top stringer (and bottom stringer, if applicable).

Unroll the mesh toward the other terminal post, and then stretch the mesh gently with a come-along (see step 8, page 344). Secure the end of the mesh to the post with a tension bar and staples, as before. Remove any excess mesh by unwinding a strand (see page 344).

Attach the bottom edge of the mesh to the tension wire every 2', using hog rings tightened with hog ring pliers. Staple the mesh to the stringers every 2' and to the line posts every 12".

Resources

AEE Solar
800-777-6609
www.aeesolar.com

Applied Energy Innovations
612-532-0384
www.appliedenergyinnovations.com
Page 256

Atkinson Electronics
800-261-3602
atkinsonelectronics.com

The Barefoot Beekeeper
www.biobees.com
Page 46

Wind Powering America
www.windpoweringamerica.gov

Credits

Crystal Liepa: 104 (top), 292 (both)

Shutterstock: 6 (top right), 8, 9 (both), 12 (both), 15 (right), 16, 21 (all), 40, 42 (both), 43 (both), 44, 100, 102, 103, 110 (top), 115, 138, 143, 149 (both), 168, 170, 172, 174, 177, 178, 180 (right), 190 (left), 204, 222 (right, both), 228, 278, 334, 340

iStock: 222 (left)

Index

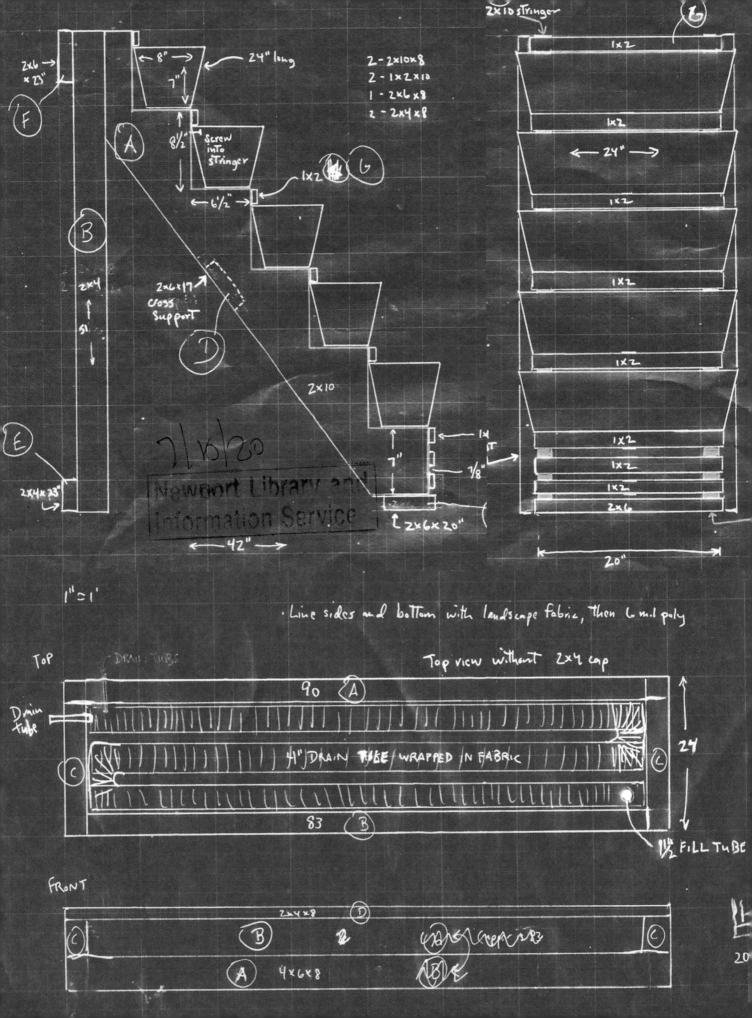

2x10 stringer   G

2x6 ×23"   F

8" → 7"   24" long

2 - 2x10x8
2 - 1x2x10
1 - 2x6x8
2 - 2x4x8

A

8½"   Screw into stringer

B   2x4   51"

1x2   G

6½"

2x6×17 cross support   D

2x10

E

2x4×23"

7"   1x   ⅞"

2x6×20"

42"

1"=1'

7/10/20

Newport Library and Information Service

1x2
1x2
24"
1x2
1x2
1x2
1x2
1x2
1x2
2x6

20"

· Line sides and bottom with landscape fabric, then 6 mil poly

Top view without 2x4 cap

TOP   DRAIN TUBE

Drain Tube

90   A

4" DRAIN TUBE WRAPPED IN FABRIC

C   83   B   C

24

1½ FILL TUBE

FRONT

2x4x8   D

B   2x4x23

C   A   4x6x8   C

20